● CONTENTS

Scotland's leading educational publishers

CfE Higher
BIOLOGY
STUDENT BOOK

J Di Mambro • A Drummond • S White

001/240615

10 9 8 7 6 5 4 3 2 1

ISBN 9780007549283

Published by
Leckie & Leckie Ltd
An imprint of HarperCollins*Publishers*
Westerhill Road, Bishopbriggs, Glasgow, G64 2QT
T: 0844 576 8126   F: 0844 576 8131
leckieandleckie@harpercollins.co.uk   www.leckieandleckie.co.uk

Special thanks to
Jouve (layout and illustration); Ink Tank (cover design);
Lee Haworth Mulvey (project management); Helen Bleck
(copy edit); Graham Moffat (proofread)

A CIP Catalogue record for this book is available from the British Library.

Acknowledgements
Whilst every effort has been made to trace the copyright holders, in cases where this has been unsuccessful, or if any have inadvertently been overlooked, the Publishers would gladly receive any information enabling them to rectify any error or omission at the first opportunity.

Leckie & Leckie would like to thank the following copyright holders for permission to reproduce their material:

SQA for Assignment support text; Figure 1.1.1 *Barrington Brown/ Science Photo Library*; Figure 1.1.1 *King's College London Archives/Science Photo Library*; Figure 1.2.3 *Henning Dalhoff/ Science Photo Library*; Figure 2.8.3 *Trevor Clifford Photography/ Science Photo Library*; Figure 3.6.3 *Biophoto Associates/Science Photo Library*; Figure 3.7.4 and 3.7.5 *NASA/Science Photo Library*; Figure 3.7.6 *Source Centre for Ecology & Hydrology, British Trust for Ornithology, Marine Biological Association and the National Biodiversity Network Gateway. (UK DEFRA)*

Printed in Italy by Grafica Veneta S.p.A.

Welcome to the CfE Higher Biology Student Book!

This book covers all of the skills, knowledge and understanding included in the CfE Higher Biology Course.

It has been designed to help you pass Higher Biology and is suitable for students who are currently following the CfE Higher course. We have assumed some prior skills and knowledge that would be gained from completing relevant Curriculum for Excellence experiences and outcomes and/ or the National 4 and 5 Biology courses.

Features

YOU SHOULD ALREADY KNOW:

You should already know:

- The nucleus of a cell contains chromosomes, which are composed of genes made of DNA.
- DNA carries genetic information for making protein.

Each chapter starts with a summary of the assumed prior knowledge. This should help to give you a background to the ideas that will be explored within the chapter.

LEARNING INTENTIONS:

Learning intentions

- Describe the structure of DNA in terms of nucleotides and complementary base pairing.

Each chapter opens with a list of topics covered in the chapter, and gives you a good idea as to what you should be able to do when you have worked your way through the chapter.

HINTS

 Hint

Remember BPS – Base, Phosphate, Sugar

MAKE THE LINK

Where appropriate, a feature that identifies links both within topics in Biology and between Biology and other subjects, is used to highlight the interconnectedness of the subject to other.

ACTIVITIES

Activity 1.1.3 Work as a group to ...

1. Design and make an A2 portrait collage to show the structure of DNA.

   You will need an A2 sheet, 6 colours of card, scissors and a glue-stick. Use a medium tip marker to label the parts. Your teacher may ask your group to present the work to your class.

These are sets of questions covering knowledge and skills and active learning tasks to work on individually, with a partner or in a group.

SUCCESS CRITERIA

Each chapter closes with a summary of learning statements showing what you should be able to do when you complete the chapter. You should use them to help identify where you are with your learning and the next steps needed to make any improvements.

COURSE ASSESSMENT

There are two components to the CfE Higher Biology Course Award:

1.	A question paper	100 marks
2.	An assignment	20 marks
	TOTAL	120 marks

COURSE ASSESSMENT STRUCTURE

COMPONENT 1 – THE QUESTION PAPER

The question paper will have two sections.

Section 1, titled 'Objective Test', will have 20 marks.

Section 2, titled 'Paper 2', will contain restricted and extended response questions and will have 80 marks.

Total of 100 marks (80% of the total mark).

Marks will be distributed approximately proportionately across the units.

The majority of the marks will be awarded for applying knowledge and understanding. The other marks will be awarded for applying scientific inquiry, scientific analytical thinking and problem-solving skills.

COMPONENT 2 – THE ASSIGNMENT

The purpose of the assignment is to allow the learner to carry out an in-depth study of a biology topic. The topic will be chosen by the learner, who will investigate/research the underlying biology and the impact on the environment/society.

The assignment will assess the application of skills of scientific inquiry and related biology knowledge and understanding.

The assignment is split into two sections:

1. Research
2. Communication

The research element of the assignment is where you will select and carry out an investigation into a relevant topic in Biology. The topic should draw from more than one of the course's key areas. It is a good idea to speak with your teacher or other assessor when making your choice of assignment topic. The following lists provide some ideas as to the assignment you might undertake. Please note that these are ideas only: try to come up with an Assignment title yourself.

Unit 1

- Case study on the bacterial transformation experiments of Griffiths.
- Case study on identification of DNA as the transforming principle by Avery *et al*.
- Case study on phage experiments of Hershey and Chase.
- Case study on Meselson and Stahl experiments on DNA replication.
- Case study on the use of PCR.
- Investigating plant evolution using chloroplast DNA and PCR.
- Tissue culture of plant material.

- Case study on use of stem cells in repair of diseased or damaged organs (e.g. skin grafts, bone marrow transplantation and cornea repair).
- Case study on ethics of stem cell research and sources of stem cells.
- Research rarity of polyploidy in animals.
- Sexual selection in brine shrimp or other organisms.
- Case study on Hybrid zones.
- Research the importance of the *Fugu* genome.

Unit 2

- Case study on the toxic effects of venoms, toxins and poisons on metabolic pathways.
- Research different use of substrates during exercise and starvation.
- Case study on adaptations to survive low-oxygen niches.
- Case study on the response of a conformer to a change in an environmental factor.
- Comparisons of marine and estuarine invertebrates and their response to variation in salinity.
- Research on aspects of surviving adverse conditions.
- Research the genetic control of migratory behaviour in studies of populations of the blackcap.

- Research different types of extremophiles.
- Research use of H_2 in methanogenic bacteria and H_2S in sulfur bacteria.
- Research industrial processes that use microorganisms. Suitable processes that involve underpinning biology include: citric acid production, glutamic acid production, penicillin production and therapeutic proteins such as insulin, human growth hormone and erythropoietin.
- Case study on bacterial transformation.
- Research the development of a microbiological product from discovery to market.

Unit 3

- Case study on challenge of providing food for the global human population. Contribution of biological science to interdisciplinary approaches to food security.
- Evaluate crop trials to draw conclusions on crop suitability, commenting on validity and reliability of trial design and the treatment of variability in results.
- Case studies on the development of particular crop cultivars and livestock breeds.
- Research self-pollinating plants - naturally inbreeding and less susceptible to inbreeding depression due to the elimination of deleterious alleles by natural selection
- Case study on the control of weeds, pests and or diseases of agricultural crops by cultural and chemical means
- Analyse data on crop planting density, biological yield and economic yield using leaf area index, crop growth rates and harvest index
- Case histories of plant mutations in breeding programmes.
- Genetic transformations in plant breeding

- Case study on the control of weeds, pests and or diseases of agricultural crops by cultural and chemical means.
- Case studies on, for example, control of glasshouse whitefly with the parasitic wasp *Encarsia*, control of glasshouse red spider mite with the predatory mite *Phytoseiulus* and/or control of butterfly caterpillars with the bacterium *Bacillus thuringiensis*.
- Investigate the chemical and biological control of red spider mite.
- Research the five freedoms for animal welfare.
- Research the links between symbioses and anthropogenic climate change.
- Case study on primate behaviour.
- Research the Permian, Cretaceous and Holocene mass extinction events.
- Analyse data on exploitation of whale or fish populations.
- Research impact of naturally low genetic diversity within cheetah populations.
- Research impact of habitat fragmentation and benefits of habitat corridors for tiger populations.
- Case study on invasive species

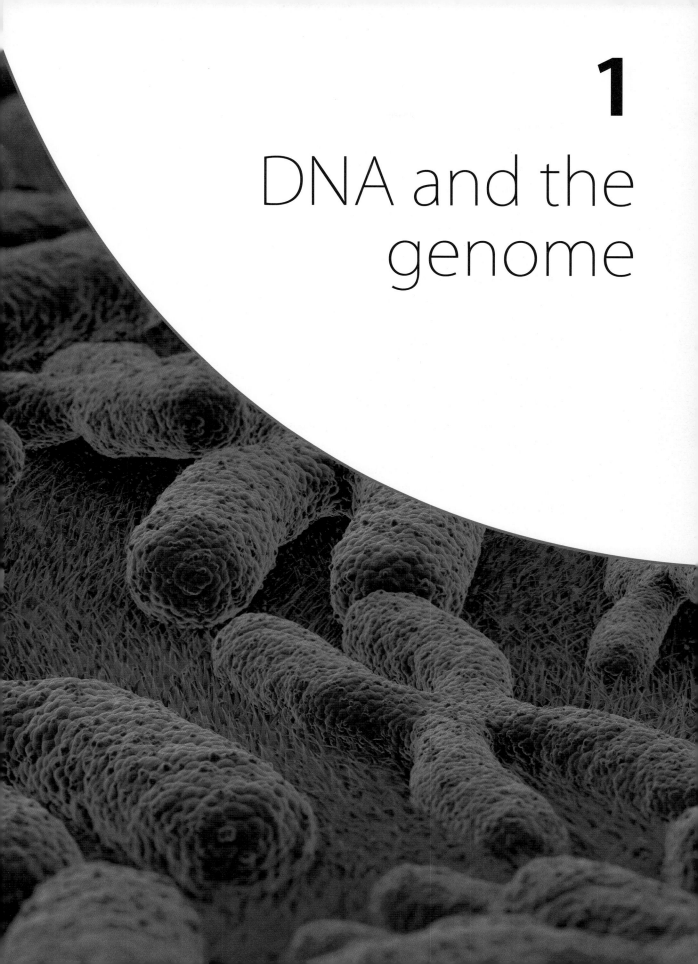

1

DNA and the genome

1.1 The structure of DNA

You should already know:

- The nucleus of a cell contains chromosomes, which are composed of genes made of DNA.
- DNA carries genetic information for making protein.
- DNA molecules are in the shape of a double-stranded helix held together by complementary base pairs.
- The four different bases in DNA are called adenine (A), guanine (G), thymine (T) and cytosine (C).
- A is always paired with T; C is always paired with G.
- The sequence of bases along one of the DNA strands makes up the genetic code.

Learning intentions

- Describe the structure of DNA in terms of nucleotides and complementary base pairing.
- Describe the components of a nucleotide.
- State how DNA strands are held together.
- Describe the antiparallel nature of the strands of DNA.
- State what is meant by the genotype of a cell.

📖 Genotype

The genetic make-up of an organism.

Introducing DNA

Deoxyribonucleic acid (DNA) is the substance that makes up the genetic material in cells and gives the cell its **genotype**.

The chemical structure of DNA was discovered in the 1950s by two scientists, James Watson and Francis Crick, working in Cambridge, UK. With Maurice Wilkins, they won the 1962 Nobel Prize in physiology and medicine for their discoveries. The evidence for their conclusions came from model-building using X-ray images of DNA crystals from the work of Rosalind Franklin and Maurice Wilkins, and others, as shown in **Figure 1.1.1**.

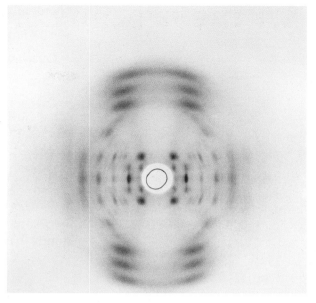

Figure 1.1.1 *James Watson (left) and Francis Crick with their model of DNA structure and an X-ray image of a DNA crystal of the type produced by Rosalind Franklin and Maurice Wilkins*

Components and structure of DNA

Nucleotides

DNA is composed of very long molecules made up of repeating chemical units called **nucleotides**.

A nucleotide has three chemical parts as shown in **Figure 1.1.2:**

- an organic base;
- a phosphate group;
- a central deoxyribose sugar.

Notice that the carbon atoms of the deoxyribose sugar are numbered. The organic base is joined to carbon 1 and the phosphate group to carbon 5.

> **⁘ Make the Link**
>
> DNA is the blueprint for life and therefore will feature throughout the course.

> **🔍 Hint**
>
> Remember **BPS** – Base, Phosphate, Sugar

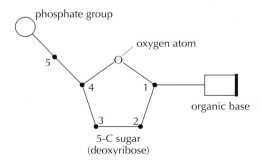

Figure 1.1.2 *A single nucleotide of DNA*

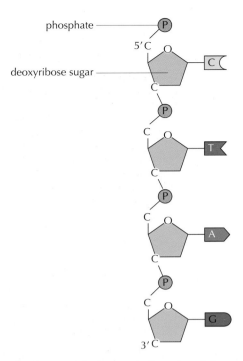

phosphate

deoxyribose sugar

Figure 1.1.3 *Short strand of DNA with four nucleotides on a 3' to 5' sugar phosphate backbone*

There are four different organic bases called:

- **Adenine** (A)
- **Guanine** (G)
- **Thymine** (T)
- **Cytosine** (C)

Nucleotides are linked together to make a long strand of DNA. This arrangement produces a sugar-phosphate backbone with the bases attached as shown in **Figure 1.1.3**.

Complementary base pairing

Two strands are joined together through weak **hydrogen bonds** that link **complementary** pairs of nucleotide bases.

> **adenine (A) always pairs with thymine (T)**
>
> **guanine (G) always pairs with cytosine (C)**

The strands run from carbon 3' (prime) at one end to carbon 5' (prime) at the other end, but in opposite directions and are described as **antiparallel**, as shown in **Figure 1.1.4**.

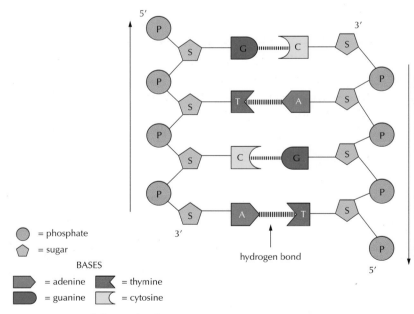

= phosphate

= sugar

BASES

= adenine   = thymine

= guanine   = cytosine

hydrogen bond

Figure 1.1.4 *Antiparallel strands of DNA*

Double helix

The double-stranded molecule is wound into a double-stranded helix as shown in **Figure 1.1.5**.

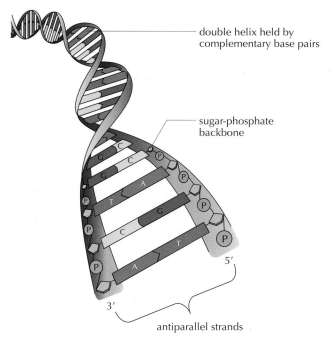

Figure 1.1.5 *Double helix of DNA showing antiparallel sugar-phosphate backbones joined by complementary base pairs*

Genotype

The genotype of a cell is its genetic information, which is coded into the sequence of bases on its DNA molecules. The information carried in base sequence of each gene is translated into a protein which can determine the physical characteristics of an organism.

Make the link

There is more about the significance of DNA structure in chapter 1.3, where you can learn about how DNA replicates itself and in chapter 1.9, where you can learn about how DNA sequences have become important in studying evolution and in chapter 2.9, where you can learn about 'sticky ends'.

Hint

✓ The key words on DNA structure are nucleotide, sugar phosphate backbone, complementary, antiparallel and double helix.
✓ A cartoon nucleotide might help!

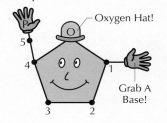

Activity 1.1.1 Work individually to ...

Restricted response

1. a) Name the **three** components which make up a nucleotide.   2
   b) Describe how the two strands of DNA are held together.   2
   c) Describe the shape of a DNA molecule.   1
   d) Explain what is meant by the following terms as applied to DNA structure:
      i. complementary
      ii. antiparallel   2
2. The table below is based on results obtained in 1952 by the scientist Erwin Chargaff, who investigated the proportions of bases in DNA samples from cells from two human sources.

(*continued*)

| Bases | Concentration in DNA sample (units) | |
	Sperm	Liver
Adenine	0·29	0·27
Guanine	0·18	0·19
Thymine	0·30	0·27
Cytosine	0·18	0·18

a) Give evidence from the table which supports the base pairing of adenine with thymine and cytosine with guanine. 1

b) Identify from the table that some experimental error affected these results. 1

3. When a DNA sample is heated, its strands separate. The heat needed to do this is proportional to the percentage of G–C base pairs as shown in the table below.

G – C base pairs (% in molecule)	Temperature needed to separate strands (°C)
0	70
20	77
40	85
60	94
80	100

On a piece of graph paper, **plot a line graph** to show the percentage of G–C base pairs against the temperature at which the strands of DNA molecules separate. 2

4. Look back at the X-ray photograph in **Figure 1.1.1**.

Try to find out what this type of image can reveal about the structure of a DNA molecule.

5. Research the work of James Watson and Francis Crick

Or Alfred Hersey and Martha Chase

Or Oswald Avery *et al*

on the internet and write a brief summary of their work on DNA. Make your piece about 150 words long; say what the researchers did and what they found out; include one diagram.

Extended response

Describe the structure of a DNA molecule. 6

GO! ## Activity 1.1.2 Work in pairs to ...

1. **Make a model** of DNA.

   Use two different colours of miniature marshmallow sweets to represent sugars and phosphates and four colours of jelly sweets for the bases. Use cocktail sticks cut in two to join the components.

   Arrange five nucleotide pairs like a ladder – don't try to model the double helix!

2. **Research** the scientists mentioned in the table below.

   Rearrange the table to match the scientists with the contributions they made to the discovery of the structure of DNA and the approximate years in which they worked.

Scientists	Contributions	Approximate date
Avery, McCarty and McLeod	Made images of DNA using X-ray crystallography techniques.	1928
Chase and Hershey	Carried out experiments on *Pneumonia* bacteria to show DNA is the genetic material of living cells.	1944
Chargaff	Used phage viruses to confirm that protein was not genetic material, so inferring that DNA could be.	1951–53
Watson and Crick	Discovered the base paring rules of DNA.	1952
Wilkins and Franklin	Discovered that there is a substance which carries genetic code.	1952
Frederick Griffiths	Suggested the double helix structure of DNA.	1953

3. **Make flash cards**.

   Write each word that is **emboldened** in the text of this chapter on a small piece of card. Write its meaning on the other side. Test each other using the cards.

GO! ## Activity 1.1.3 Work as a group to ...

1. **Design and make** an A2 collage to show the structure of DNA.

   You will need an A2 sheet, 6 colours of card, scissors and a glue-stick. Use a medium tip marker to label the parts. Your teacher may ask your group to present the work to your class.

2. **Produce a proposal** from your group as to which scientist(s) made the **most** significant contribution to the discovery of the structure of DNA.

   You will need the proposal and several supporting pieces of evidence and opinion. Debate the issue with your class and take votes!

After working on this chapter, I can:

1. Describe the genotype of a cell as its sequence of DNA bases.

2. State that each strand of DNA is made up from chemical units called nucleotides.

3. State that each nucleotide is made up of a deoxyribose sugar, a phosphate and an organic base.

4. State that deoxyribose sugar molecules have five carbon atoms which are numbered 1 to 5.

5. Identify carbon 3 (3') and carbon 5 (5') of a deoxyribose sugar molecule.

6. State that nucleotides on the same strand are joined to form a sugar-phosphate backbone.

7. State that a DNA strand has a deoxyribose at its 3' end and a phosphate group at its 5' end.

8. State that the nucleotides of one strand of DNA are linked to the nucleotides on the second strand through their bases – the bases form pairs joining the strands.

9. State that the bases pair in a complementary way; adenine always pairs with thymine and guanine always pairs with cytosine.

10. State that base pairs are held together by hydrogen bonds.

11. Explain that the two strands of a DNA molecule run in opposite directions and are described as antiparallel to each other.

1.2 The organisation of DNA

You should already know:

- The ultrastructure of typical plant, animal, fungal and bacterial cells.
- Bacterial cells contain additional DNA in structures called plasmids.

Learning intentions

- State that circular chromosomal DNA and plasmids are found in prokaryotes.
- State that circular plasmids are found in yeast cells.
- State that circular chromosomes are located in mitochondria and chloroplasts of eukaryotes.
- State that DNA in eukaryotes is found in linear chromosomes.
- Describe these linear chromosomes as tightly coiled and packaged with associated proteins.

Prokaryotes and eukaryotes

Prokaryotes

Cells which lack a true membrane-bound nucleus, such as bacteria, are called **prokaryotes.**

Due to the lack of a true nucleus, prokaryotic genetic information is contained in a large **circular chromosome**, located in the cytoplasm.

In addition to this, many prokaryotes have additional small circular DNA structures called **plasmids**.

These structural features are displayed in **Figure 1.2.1**.

> 📖 **Prokaryote**
>
> Organism with a single cell which lacks a true membrane-bound nucleus.

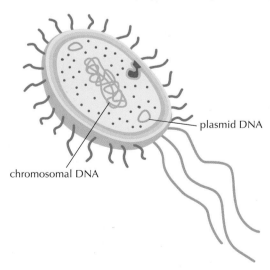

plasmid DNA

chromosomal DNA

Figure 1.2.1 *Structure of a bacterial cell (prokaryote)*

Eukaryote

Organism made of cells, which contains a true membrane-bound nucleus.

Organelle

A sub-cellular structure within a cell such as a mitochondrion or vacuole that performs a specific function.

Eukaryotes

Cells with a true nucleus and other membrane-bound organelles are **eukaryotic**. Eukaryotic organisms include yeasts, plants and animals.

Linear chromosomes, located in the nucleus, contain much of a eukaryotic cell's DNA. However, some of the DNA in eukaryotes is found in circular chromosomes, located in two structures, the mitochondria of all eukaryotic cells and the chloroplasts of green plant cells, as shown in **Figure 1.2.2**.

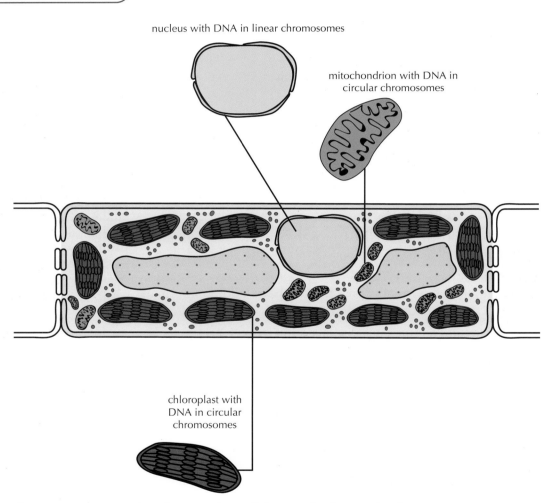

nucleus with DNA in linear chromosomes

mitochondrion with DNA in circular chromosomes

chloroplast with DNA in circular chromosomes

Figure 1.2.2 *Structure of a eukaryotic plant cell showing distribution of DNA*

The length of DNA in a cell is many times longer than the cell itself, so it has to be arranged in a special way in order to fit. Therefore, DNA in the linear chromosomes in the nucleus of eukaryotes is tightly coiled and packaged with associated proteins. These can assemble and unwind when required to do so.

Hint

Remember: yeast cells are unusual as they are eukaryotes that contain plasmids.

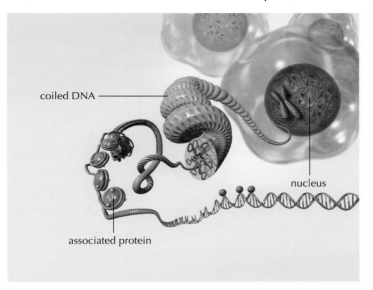

coiled DNA

nucleus

associated protein

Figure 1.2.3 *A linear chromosome with tightly coiled DNA packaged with associated proteins*

Yeast (fungal) cells are eukaryotic and unicellular, and they also contain circular plasmids.

The table below summarises the organisation of DNA in eukaryotic and prokaryotic cells.

Hint

Memorise the key differences between eukaryotes and prokaryotes as shown in the table on the this page.

Make the link

The structure of DNA is discussed in chapter 1.1, while details on the replication of DNA and its importance follow in chapter 2.1. This includes the polymerase chain reaction, a technique which creates many copies of a strand of DNA.

Feature	Cell type	
	Eukaryotic	**Prokaryotic**
True membrane bound nucleus?	Yes	No
Chromosome structure	Linear and circular	Circular
Location of genetic material	Nucleus, chloroplast, mitochondria (and plasmids in yeast)	Cytoplasm and plasmids
Examples	Plants, animals and yeasts	Bacteria

GO! Activity 1.2.1 Work individually to ...

Restricted response

1. **State** the form into which DNA is organised in the nucleus of a eukaryotic cell. 1
2. **Give one example** of the following types of cells:
   a) Eukaryote
   b) Prokaryote. 2
3. State **two** structures in eukaryotic cells which contain circular chromosomes. 2
4. **Describe** how a 4-metre strand of DNA is organised to enable it to fit into a cell's microscopic nucleus. 2

Extended response

Describe the differences between prokaryotic and eukaryotic cells. 6

GO! Activity 1.2.2 Work in pairs to ...

1. **Research** the different types of eukaryotic and prokaryotic cells, making a list of their names, location and arrangements of DNA within their cells.
2. **Make flash cards**, highlighting the differences between a eukaryotic and prokaryotic cell, including examples, location of DNA, structural differences and arrangement of DNA.

GO! Activity 1.2.3 Work as a group to ...

Create a poster contrasting the differences between eukaryotic and prokaryotic cells.

Include a labelled diagram of an animal, plant, yeast and bacterial cell, highlighting the areas where DNA is found and stating the form into which it is arranged. This could be presented to the class.

After working on this chapter, I can:

1. State that DNA is arranged in circular chromosomes and plasmids in prokaryotes.

2. State that DNA is arranged as circular chromosomes and plasmids in prokaryotic cells.

3. State that DNA is arranged in linear chromosomes in the nuclei of eukaryotic cells.

4. State that some DNA is arranged in circular plasmids in yeast.

5. State that circular chromosomes are located in the mitochondria in all eukaryoic cells.

6. State that circular chromosomes are located in the chloroplasts of eukaryotic green plant cells.

7. Explain that DNA in linear chromosomes is tightly coiled and packaged with associated proteins in the nucleus of eukaryotes.

1.3 Replication of DNA

Learning intentions

- DNA copies itself by replication.
- Describe the roles of DNA polymerase and a primer in DNA replication.
- Explain directionality of replication on both template strands.
- State that DNA polymerase adds complementary nucleotides to the 3′ end of a DNA strand.
- State that fragments of DNA are joined together by ligase.
- Describe the stages in the polymerase chain reaction (PCR) amplification of DNA using complementary primers for specific target sequences.
- State the role of endonuclease in PCR.

DNA replication

It is essential that DNA is able to replicate (make more exact copies of itself) so that all cells have enough genetic information to be able to divide by mitosis. The basis for this process is the complementary base paring rule you learned about in National 5:

- adenine pairs with thymine
- cytosine pairs with guanine

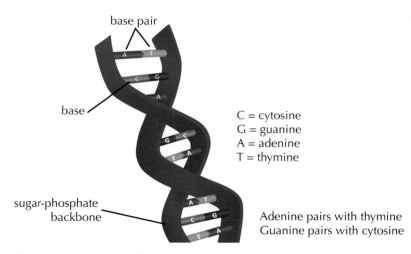

C = cytosine
G = guanine
A = adenine
T = thymine

Adenine pairs with thymine
Guanine pairs with cytosine

base pair

base

sugar-phosphate backbone

Figure 1.3.1 *Structure of DNA*

This is shown in **Figure 1.3.1**. In order for DNA replication to occur, certain substances must be present within the nucleus:

- **DNA template** (parental strands – to allow an exact copy to be made)
- **free DNA nucleotide bases** (all four types)
- **enzymes** (such as DNA polymerase and ligase)
- **primers** (their presence is required to start replication)
- **ATP** (to supply energy for the process)

Main stages in the process

DNA replication involves the following stages, listed below, and in **Figure 1.3.2**.

Stage 1 – separation of two parental DNA strands (templates)

Stage 2 – free DNA nucleotides pair with complementary bases on template strands

Stage 3 – sugar-phosphate backbones form on new strands

Stage 4 – two daughter DNA molecules formed which are genetically identical to parental DNA

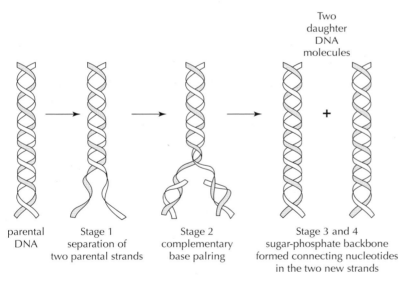

	Two daughter DNA molecules
parental DNA	Stage 1 separation of two parental strands	Stage 2 complementary base pairing	Stage 3 and 4 sugar-phosphate backbone formed connecting nucleotides in the two new strands

Figure 1.3.2 *Basic DNA replication steps*

During stage 1, the original parental DNA is separated using an enzyme and ATP to break **weak hydrogen bonds** between base pairs. The strands then unwind and unzip to form two template strands. This happens at several points along the DNA molecule.

In stage 2, a **primer** is attached to the 3′ end of the parental DNA strand, then an enzyme called **DNA polymerase** begins to join free DNA nucleotides to the template strands.

📖 Weak hydrogen bonds

Bonds between the base pairs which hold both sides of a DNA molecule together.

📖 DNA polymerase

Enzyme which adds complementary nucleotides to the deoxyribose (3′) end of a DNA strand.

📖 Primer

Short, single strand of DNA bases required for replication to begin.

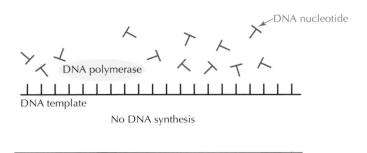

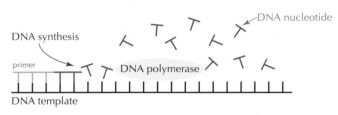

Figure 1.3.3 *Role of a primer in DNA replication*

Leading strand

On the leading strand, nucleotides are added from the deoxyribose (3′) end of the parental strand in one direction in a continuous fashion.

Lagging strand

The lagging strand is antiparallel to the leading strand but replication cannot begin at the 5′ end. Instead, it is replicated in fragments, which starts with primer at the 3′ end and adds nucleotides in a discontinuous fashion as more primer is added to the replication fork as shown in **Figure 1.3.4**.

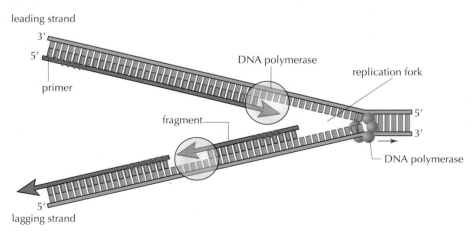

Figure 1.3.4 *Leading and lagging strand replication*

📖 Ligase

Enzyme which joins fragments of DNA together on the lagging strand.

Stage 3 involves the use of the enzyme **ligase**, which joins the small fragments of newly formed DNA in the lagging strand together to form a sugar phosphate backbone.

In order for an organism to produce new cells and to pass on the correct genetic instructions to new generations, DNA replication must create identical copies of the information contained in the DNA. This is made possible by the exact replication of sequences of the bases which make species distinct from others and allow **specific proteins to be synthesised** using the genes/specific base sequences on the DNA.

Polymerase chain reaction (PCR)

The **polymerase chain reaction** (PCR) is a technique for the **amplification** of DNA *in vitro*.

In PCR, the chosen primers are complementary to specific target sequences at the two ends of the region of DNA to be amplified. This will often involve the use of a **thermal cycler**.

Similar to DNA replication, PCR has some initial requirements for the process to occur. These are:

- DNA template
- free DNA nucleotides (all four types)
- heat-tolerant DNA polymerase (enzyme)
- primers – artificially made, short, single strands of DNA that use bases complementary to those at either end of the DNA fragment to be copied.

Figure 1.3.6 shows the general stages of the PCR cycle.

Figure 1.3.5 *PCR thermal cycler*

📖 **Polymerase chain reaction (PCR)**

In vitro method of amplifying a sequence of DNA.

📖 **Amplification**

Create many copies of a fragment of DNA.

📖 **In vitro**

In an artificial environment outside a living organism.

📖 **Thermal cycler**

Automated machine able to carry out repeated cycles of PCR by varying temperature.

🔍 **Hint**

The number of DNA molecules doubles at each cycle.

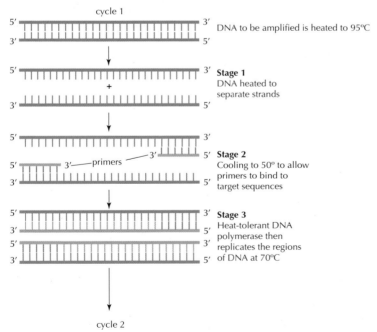

cycle 1

5′   3′   DNA to be amplified is heated to 95°C
3′   5′

5′   3′ **Stage 1**
+   DNA heated to separate strands
3′   5′

5′   3′
3′ — primers — 3′ 5′ **Stage 2**
5′   Cooling to 50° to allow primers to bind to target sequences
3′   5′

5′   3′ **Stage 3**
3′   5′ Heat-tolerant DNA polymerase then replicates the regions of DNA at 70°C
5′   3′
3′   5′

cycle 2

Figure 1.3.6 *The three stages of PCR*

The stages

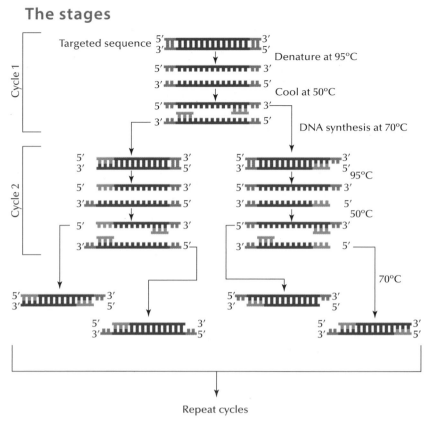

Figure 1.3.8 *Repeated cycles of PCR amplifying the original DNA strand*

1. DNA heated to 95°C to separate strands, then cooled to 50°C to for **primer** binding.

2. Cooling allows primers to bind to target sequences.

3. Heat-tolerant **DNA polymerase** then replicates the region of DNA at 70°C.

4. Repeated cycles of heating and cooling **amplify** this region of DNA as shown in **Figure 1.3.8**.

Figure 1.3.9 shows the stages and temperatures.

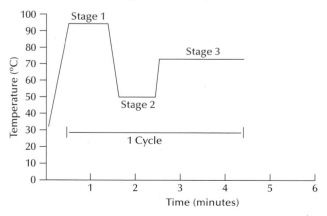

Figure 1.3.9 *The stages of PCR in a time vs. temperature graph*

When a thermal cycler is used, the PCR cycle can be repeated many times, as shown in **Figure 1.3.10**.

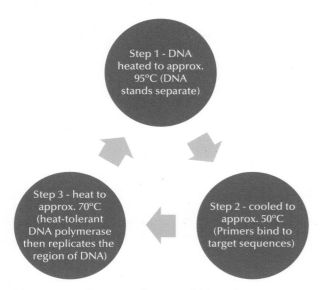

Figure 1.3.10 *Diagram of steps in PCR cycle*

🔍 Hint

Memorise the stages of PCR, especially the approximate temperatures at which each stage occurs.

⁘ Make the link

There is more on taxonomic groups in Chapter 1.9.

Practical applications of PCR

PCR can be used in many ways, for example:

- DNA sequencing (creating many copies of a small piece of DNA can allow it to be fully sequenced).

- genetic mapping studies (for example, the Human Genome Project)

- forensic and parental testing (this is often critical for forensic analysis, when only a trace amount of DNA is available as evidence).

- sex determination in pre-natal cells (sex can be determined with the help of presence of a unique sequence on the y chromosome (SRY gene). If this gene is present it is male, if it is absent it is female).

- classification of species into taxonomic groups based on DNA sequences (small fragments of DNA can be amplified to allow genomes to be compared).

- screening for and diagnosis of genetic disorders (such as cystic fibrosis)

🔍 Hint

For all biological processes, such as DNA replication and PCR, always ensure you learn all of the steps involved, particularly details such as temperatures, locations and structures, and ensure you can order them correctly.

GO! Activity 1.3.1 Work individually to ...

Restricted response

1. State **four** different molecules required for DNA replication to occur. 2
2. Describe the function of the following in DNA replication:
   a) Primer
   b) DNA polymerase
   c) Ligase 3
3. Describe the key differences in replication on the leading and lagging strands of DNA. 2
4. State the primary purpose of PCR. 1
5. **Give** two practical application of PCR. 2

Extended response

Describe the main steps involved in
- DNA replication 5
- The polymerase chain reaction (PCR) 4

GO! Activity 1.3.2 Work in pairs to ...

Carry out a **case study** on the uses of PCR. Use the examples given in the chapter to get you started. You should try to include experimental data in your case study and try to process this as practice for your assignment. You may also wish to evaluate the data you find as well as draw a conclusion based on what it shows.

GO! Activity 1.3.3 Work as a group to ...

Produce a time lapse film, using plasticine or plastic models, to show the stages in either DNA replication or PCR. Present it to the class. Take a vote on the best film and discuss.

After working on this chapter, I can:

1. State the basic requirements for DNA replication – DNA template, free DNA nucleotide bases, enzymes, primers, ATP.

2. State that the function of DNA polymerase is to join free nucleotide bases to the template DNA strand.

3. State that the function and of ligase is to join fragments of DNA together on the lagging strand.

4. Describe the role of primers as small sections of DNA required for replication to begin.

5. Explain replication on the leading strand as continuous and starting at the 3′ of the parental strand, and describe the lagging strand as discontinuous and replicated in fragments.

6. Describe the main stages in DNA replication as –

    Stage 1 – separation of two parental DNA strands (template)

    Stage 2 – free DNA nucleotides pair with complementary bases on template strands

    Stage 3 – sugar phosphate backbones form on new strands

    Stage 4 – two daughter DNA molecules formed which are genetically identical to parental DNA.

7. State the basic requirements for PCR as DNA template, free DNA nucleotides (all four types), heat-tolerant DNA polymerase, primers.

8. Describe the polymerase chain reaction (PCR) as a technique for the amplification of DNA *in vitro*.

9. Describe the main stages in PCR as –

    Stage 1 – DNA is heated to separate strands.

    Stage 2 – Cooling allows primers to bind to target sequences.

    Stage 3 – Heat-tolerant DNA polymerase replicates the DNA.

1.4 Control of gene expression

You should already know:

- The sequence of bases in a DNA molecule determines the amino acid sequence in protein.
- Messenger RNA (mRNA) is a molecule which carries a complementary copy of the DNA code.
- mRNA moves from the nucleus to a ribosome where protein is assembled from amino acids.
- The variety of protein shapes and functions arises from their sequences of amino acids.
- Protein functions include structural units, enzymes, hormones, antibodies and receptors.

Learning intentions

- State what is meant by 'phenotype'.
- Compare the structure of RNA and DNA.
- Describe the structure and functions of mRNA.
- Describe the process of transcription.
- Describe the process of translation.
- Explain that gene expression is controlled through the regulation of transcription and translation.
- Explain how different proteins can be expressed from one gene.
- Describe post-translation protein structure modification.
- Describe the three-dimensional shape and structure of proteins.

📖 Gene expression

Process involving transcription and translation where DNA sequences are used to direct the production of proteins.

📖 Transcription

Making a primary transcript of mRNA using a DNA sequence Takes place in the nucleus of a cell.

📖 Translation

Production of a polypeptide chain informed by an mRNA sequence. Takes place in a ribosome.

Gene expression

Gene expression is controlled by the regulation of **transcription** and **translation**. The genetic code processed by transcription and translation is found in all forms of life.

An organism's **phenotype** is determined by the proteins produced as the result of gene expression. This can be influenced by intra- and extra-cellular environmental factors. All of an organism's DNA sequences are called its genome. Genes make up only a part of the genome and not all genes are expressed in every cell in an organism.

Ribonucleic acid (RNA)

The previous chapter discussed a molecule called DNA. Another molecule, vital to the process of protein synthesis, is **ribonucleic Acid (RNA)**. It has some similarities to DNA, but differs in three main ways:

- RNA is a **single-stranded** molecule
- RNA has the same bases as DNA, except for **uracil**, which replaces thymine
- RNA has a **ribose sugar**, in contrast to DNA's deoxyribose sugar, as shown in **Figure 1.4.1**.

See the table for a comparison of DNA and RNA.

Feature	DNA	RNA
Number of strands	2	1
Bases	A T G C	A U G C
Sugar	Deoxyribose	Ribose

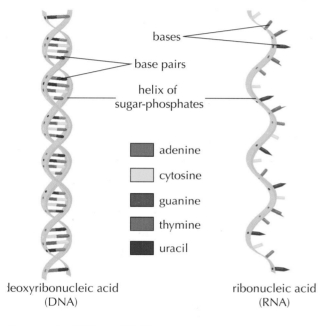

bases
base pairs
helix of sugar-phosphates

- adenine
- cytosine
- guanine
- thymine
- uracil

deoxyribonucleic acid (DNA)

ribonucleic acid (RNA)

Figure 1.4.1 *DNA and RNA*

Figure 1.4.3 shows an overview of the stages involved in gene expression. This is useful in gaining an overall picture of the entire process.

Three main forms of RNA are:

- **Messenger RNA** (mRNA) carries a copy of the DNA code from the nucleus to the ribosome (see **Figure 1.4.3**). It has a linear form and groups of three bases known as **codons**.
- **Ribosomal RNA** (rRNA), along with ribosomal protein, form protein-synthesising organelles called **ribosomes**.
- **Transfer RNA** (tRNA) molecules each carry a specific amino acid and are involved in the second part of protein synthesis. They have a folded shape and an **anticodon**,

📖 **Ribonucleic acid (RNA)**

A molecule similar to DNA and is essential for protein synthesis.

📖 **Phenotype**

Genetically determined characteristics of an organism.

✷ **Make the link**

There is more about genomes in Chapter 1.6.

📖 **Codon**

Triplet of bases in mRNA that codes for a specific amino acid which is carried by tRNA.

📖 **Ribosome**

Site of protein synthesis. Composed of ribosomal protein and rRNA.

📖 **Anticodon**

Triplet of bases in tRNA that codes for a specific amino acid, and is complementary to a specific codon in mRNA.

which is a group of three bases, each of which attaches to a different amino acid at an attachment site, shown in **Figure 1.4.2**.

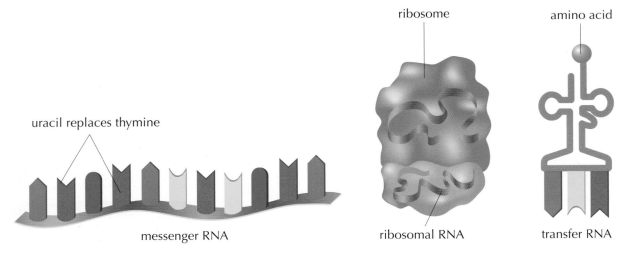

Figure 1.4.2 *Types of RNA*

Protein synthesis

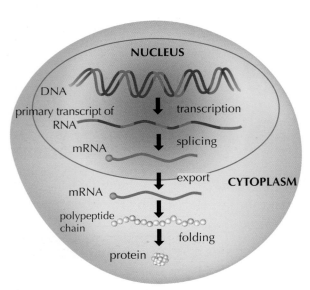

Figure 1.4.3 *Gene expression overview*

📖 **RNA polymerase**

Enzyme which unwinds DNA during transcription. Also adds free nucleotides to a single strand of DNA to form a single strand of mRNA.

Protein synthesis is a process in which instructions from DNA sequences are carried to ribosomes and proteins are synthesised. (See **Figure 1.4.3**.)

Transcription and splicing

The DNA sequence for the gene to be expressed is copied onto a primary transcript of mRNA, using complementary base pairing and RNA polymerase, in a process with the following stages:

Transcription

1. **RNA polymerase** moves along the DNA and unwinds the double helix.

2. Hydrogen bonds between base pairs break, unzipping the double helix.

3. As RNA polymerase breaks the bonds, it synthesises a primary transcript of mRNA on the DNA template strand using free RNA nucleotides. These RNA nucleotides form hydrogen bonds with the exposed DNA bases by complementary base pairing.

4. A primary transcript is formed.

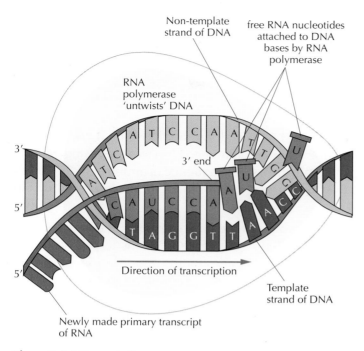

Figure 1.4.4 *Transcription*

Splicing

The primary transcript is made of **introns** and **exons**.

5. The introns of the primary transcript of mRNA are **non-coding** and are removed.

6. The exons are **coding** regions and are joined together to form mature transcript.

7. This process is called **RNA splicing**.

The process is summarised in **Figure 1.4.5**.

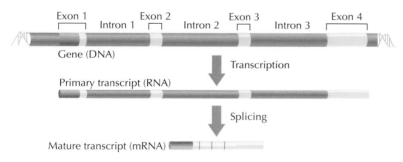

Figure 1.4.5 *RNA splicing from primary to mature transcript*

After leaving the nucleus through a pore in the nuclear membrane, the mature transcript travels through the cytoplasm to a ribosome for the next stage of protein synthesis, as shown in **Figure 1.4.4**.

📖 **Introns**

Non-coding sequences of DNA.

📖 **Exons**

Coding sequences of DNA.

📖 **Non-coding sequence**

Sequence of DNA bases that do not enter transcription and translation. No protein is synthesised.

📖 **Coding sequence**

Sequence of DNA bases that enters transcription and translation, resulting in a finished protein.

📖 **RNA splicing**

Removal of introns and joining of exons to form a mature transcript.

Peptide bonds

Chemical bonds which join amino acids together to form a polypeptide chain.

Polypeptide

Chain of many amino acids.

Start codon

First triplet which codes for the start of a polypeptide chain formation, e.g. AUG.

Stop codon

Final triplet which codes for termination in polypeptide chain formation, e.g. UAA.

Translation

The mature transcript of mRNA arrives and attaches itself to a site on the ribosome. The sequence of triplets of bases (codons) on the mRNA are then read as the complementary tRNA triplets (anticodons) carrying the appropriate amino acids are brought to the ribosome.

As the mRNA strand moves along the ribosome it continues to be read and amino acids continue to bind together using **peptide bonds**. As the chain grows, the sequence of amino acids becomes a **polypeptide**. tRNA then exits from the ribosome, as shown in **Figure 1.4.6**.

The start and stop points of the chain formed are determined by triplets of mRNA called start and stop codons respectively. The mRNA codon AUG which codes for the amino acid methionine (met) also acts as a **start codon**. The mRNA codons UAA, UGA and UAG do not code for a specific amino acid, instead acting as **stop codons** to terminate the polypeptide chain formation.

tRNA can carry out its function as it folds due to base pairing to form a triplet anticodon site and an attachment site for a specific amino acid.

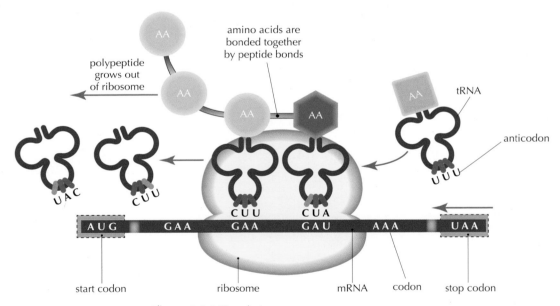

Figure 1.4.6 *Translation*

Translation can be summarised in the following steps:

1. mRNA attaches to site on the ribosome.

2. The first codon of the mRNA is called a start codon and this starts translation.

3. Codons match complementary bases on the anticodon of tRNA which transports appropriate amino acids to the ribosome. The anticodon on a tRNA molecule is specific to only one amino acid.

4. The amino acids form a chain, joined by peptide bonds.

5. This chain is called a polypeptide.

6. The last codon of the mRNA is called a stop codon and this stops translation.

7. The polypeptide leaves the ribosome once completed.

8. The way in which polypeptide chains are assembled determines the structure and function of the finished protein.

Proteins

Different proteins can be expressed from one gene, as a result of alternative RNA splicing and **post-translational modification**.

Different mRNA molecules are produced from the same primary transcript depending on which RNA segments are treated as exons and introns.

Even after translation, some proteins require subsequent modification to allow them to perform their functions.

Post-translational protein structure modification can be achieved in a number of ways, for example, by cutting and combining polypeptide chains or by adding phosphate or carbohydrate groups to the protein, shown in **Figure 1.4.8**. Insulin is synthesised after extensive post-translational modification. The original protein produced in translation undergoes three modifications before it becomes the protein desired.

> 📖 **Post-translational modification**
>
> Alterations to polypeptide chains following translation, such as the addition of a non-protein molecule, e.g. the addition of an iron atom in the blood protein haemoglobin.

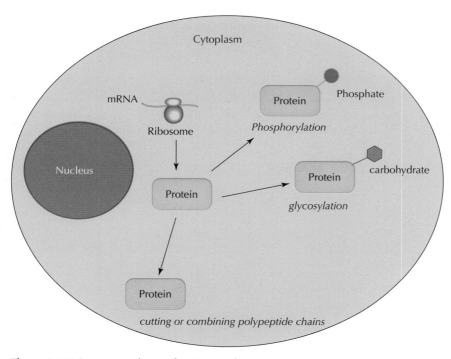

Figure 1.4.7 *Protein synthesis of amino acids*

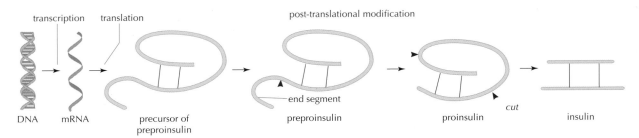

Figure 1.4.8 *Post-translational modification of insulin*

Protein structure

The proteins produced after translation and modification have a wide variety of structures and functions. The polypeptide chains, composed of amino acids joined by peptide bonds, are sequenced in a certain order which is specific to the protein function. They are also shaped in a particular way depending on their amino acid sequence.

> Proteins are composed of four main elements:
>
> carbon (C), hydrogen (H), oxygen (O), nitrogen (N).
>
> A few contain sulfur (S).

The amino acids which make up the proteins are of 20 different types.

Protein shape variation arises from folding to create a 3-dimensional shape. Folding of the polypeptide chains happens due to linking **hydrogen bonds** and other interactions between amino acids.

📖 Hydrogen bond

Weak link between base pairs in DNA. They also join amino acids at different locations in the folding of polypeptide chains to form proteins.

⁙ Make the link

Complementary base pairing was discussed in Chapter 1.1 and is important for transcription in protein synthesis.

🔎 Hint

Remember this sequence: BAP = Base → Amino Acid → Protein. Transcription comes before translation. Remember this by thinking that before you can translate you need a script to read from.

GO! Activity 1.4.1 Work individually to …

Restricted response

1. **Describe** what is meant by 'gene expression'.   1
2. **Name** the main enzyme involved in transcription.   1
3. **State** the location of transcription within a cell.   1
4. **Name** the coding and non-coding regions of mRNA.   2
5. **Describe** RNA splicing of a primary mRNA transcript.   2
6. **State** the location of translation within a cell.   1
7. **State** the name given to the triplet of bases on tRNA.   1
8. **Name** the bonds which join amino acids into a chain.   1
9. **Describe** possible modifications made to protein structure after translation.   2

Extended response

10. **Describe** the main types of protein structure.   3
11. **State three structural differences** between the structure of DNA and RNA.   3
12. **State three different types of RNA** and describe the differences in their
    structure and function.   3
13. **Describe protein synthesis** under the following headings:
    a) Transcription   5
    b) Translation   5

GO! Activity 1.4.2 Work in pairs to …

Research the variety of functions and structures of protein groups such as enzymes, antibodies and hormones.
Online, use computer software such as *Jmol* or *RasMol* to look at protein structure visualisations.

GO! Activity 1.4.3 Work as a group to …

Produce a still-frame video showing the steps involved in protein synthesis. Plasticine or DNA modelling kits are useful to show the process, as are drawn sketches. Add a narration or text captions to your video.

After working on this chapter, I can:

1. Explain that the phenotype is determined by the proteins produced as the result of gene expression, influenced by intra- and extra-cellular environmental factors.

2. State that only a fraction of the genes in a cell are expressed.

3. Describe the structure and functions of RNA.

4. State that rRNA (ribosomal RNA) and proteins form the ribosome.

5. State that each tRNA (transfer RNA) carries a specific amino acid.

6. Describe mRNA as a messenger molecule which carries a complementary copy of the DNA code from the nucleus to the ribosome.

7. Describe transcription of DNA into primary and mature RNA transcripts including the role of RNA polymerase and complementary base pairing.

8. State that introns of the primary transcript of mRNA are non-coding and are removed in RNA splicing.

9. State that exons are coding regions and are joined together to form a mature transcript. This process is called RNA splicing.

10. Describe translation of mRNA into a polypeptide by tRNA at the ribosome.

11. Describe how tRNA folds due to base pairing to form a triplet anti codon site and an attachment site for a specific amino acid.

12. State that triplet codons on mRNA and anticodons translate the genetic code into a sequence of amino acids.

13. State that start and stop codons exist.

14. Describe codon recognition of incoming tRNA, peptide bond formation and exit of tRNA from the ribosome as polypeptide is formed.

15. State that different proteins can be expressed from one gene as a result of alternative RNA splicing and post-translational modification.

16. Explain that different mRNA molecules are produced from the same primary transcript depending on which RNA segments are treated as exons and introns.

17. State that post-translational protein structure modification can be achieved by cutting and combining polypeptide chains or by adding phosphate or carbohydrate groups to the protein.

18. State that proteins are held in a 3-dimensional shape by peptide bonds, folded polypeptide chains, hydrogen bonds and other interactions between individual amino acids.

1.5 Cellular differentiation

You should already know:

- Stem cells in animals can divide and have the potential to become different types of cell.
- Stem cells are involved in growth and repair.
- Meristems are the sites of production of non-specialised cells in plants and are the sites for mitosis in a plant.
- Cells produced in meristems have the potential to become other types of plant cell and they contribute to plant growth.

Learning intentions

- Describe the general process of cellular differentiation.
- Explain the terms specialised, unspecialised and differentiated.
- Describe differentiation in plants cells.
- Describe stem cells in animals.
- Describe the main uses of stem cells for research and therapy.
- Discuss and state the ethical issues involving stem cell use and regulation.

Differentiation in plants and animals

Cellular differentiation is the process by which a cell develops more specialised functions by switching on those genes linked to the characteristics of that type of cell. For example, a stem cell can differentiate into a skin cell by switching on certain genes, which code for particular proteins, and switching off the others.

Meristems

In plants, differentiation into **specialised** cells occurs in sites called **meristems**. A meristem contains cells which are **unspecialised** and will go on to either specialise or divide to make more meristematic cells. This gives them the ability to regenerate damaged or eaten parts. For example, grasses with low meristems can grow after being grazed upon.

Growth of a plant originates from the apical and lateral meristems. The apical meristems are located in the shoot and root tips, as shown in **Figure 1.5.1**. They are responsible for the primary growth of the plant. The lateral meristems are responsible for the secondary growth, or thickening, of the plant stem. Examples of specialised plant cells include xylem, which form tubes to carry water; and phloem, which form tubes to carry food.

> ### 📖 Specialised
> Differentiated cells with a specific function and unable to become any other kind of cell.

> ### 📖 Meristems
> Regions of unspecialised cells in plants that are capable of cell division.

> ### 📖 Unspecialised
> Undifferentiated cells capable of becoming any kind of cell.

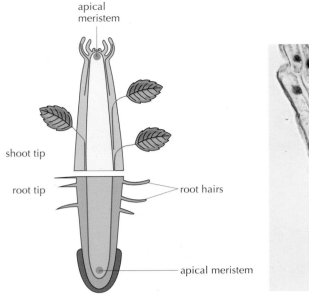

Figure 1.5.1 *Apical meristem sites*

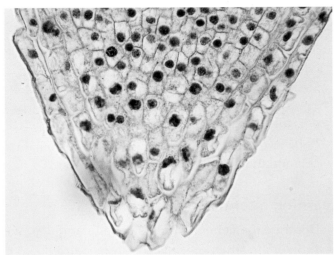

Figure 1.5.2 *Meristematic cells*

📖 Stem cells

Unspecialised somatic cells in animals that can divide to make copies of themselves (self-renew) and/or differentiate into specialised cells.

📖 Somatic

Any cell within an animal which is not a gamete (egg / sperm).

📖 Differentiate

When a cell becomes a specific, more specialised type of cell through gene expression.

Stem cells

Stem cells are relatively unspecialised cells in animals that can continue to divide and can **differentiate** into specialised cells. They become specialised after differentiation when specific genes are expressed (switched on) while others are not (switched off). This helps a wide variety of cells within animals to be differentiated from unspecialised stem cells, as shown in **Figure 1.5.3**.

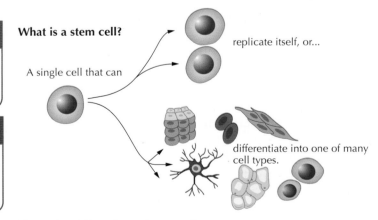

What is a stem cell?

A single cell that can

replicate itself, or...

differentiate into one of many cell types.

Figure 1.5.3 *Functions of stem cells*

Embryonic and tissue (adult) stem cells

Stem cells can be found in two main sources: in embryos and in tissues of adult animals.

Embryonic stem cells are removed from an early embryo (blastocyst) and are capable of differentiating into any type of cell (pluripotent) at this stage. These cells don't self-renew *in vivo*, but can under the right conditions in the laboratory. It is then that they have great therapeutic importance, so are used as a source of stem cells in research and medical therapy, as shown in **Figure 1.5.4**.

> ## 📖 Embryonic stem cells
> Stem cells removed from an early embryo.

> ## 📖 in vivo
> Within a living organism

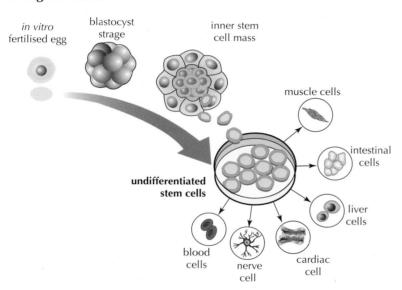

Figure 1.5.4 *Embryonic stem cell isolation and culture*

Tissue stem cells are removed from adult red bone marrow and are also capable of differentiating into many different cell types. Tissue (adult) stem cells are needed for growth, repair and renewal of tissues. They replenish differentiated cells that need to be replaced and give rise to a more limited range of cell types, e.g. blood stem cells found in the bone marrow produce the various blood cell types. The therapeutic uses of these cells are valuable and skin, blood and nerve cells can all be derived from tissue stem cells.

> ## 📖 Tissue stem cells
> Stem cells removed from adult tissue.

Research and therapeutic uses of stem cells

Stem cells have been widely researched during recent years and are now becoming important for **therapeutic uses** such as the repair of damaged or diseased organs or tissues, e.g. corneal transplants and skin grafts for burns.

> ## 📖 Therapeutic uses
> Testing part of a medical therapy.

📖 Drug testing

Testing potential medications to treat disease.

Stem cell research provides information on how cell processes such as cell growth, differentiation and gene regulation work.

Stem cells can be used as models to study how diseases develop, or for **drug testing**, as well as for therapeutic uses. For example, leukaemia is a cancer of the blood caused by uncontrolled division of white blood cells. This can be treated by destroying the cancerous cells and using stem cells derived from bone marrow in a transplant to replace those destroyed, as shown in **Figure 1.5.5**.

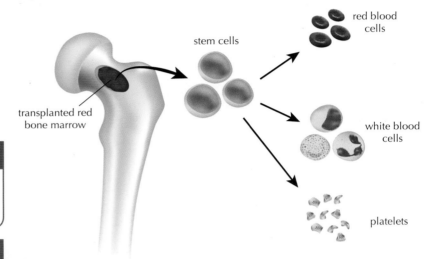

Figure 1.5.5 *Tissue stem cells used in blood cancer treatment*

📖 Ethical issues

Issues arising from a set of principles of right conduct and moral values.

📖 Induced pluripotent stem cells

Specialised cells which are reprogrammed to an embryonic state and have the potential to become a different type of cell.

📖 Nuclear transfer techniques

The introduction of the nucleus from a cell into an enucleated egg cell. The donor nucleus used for nuclear transfer may come from a differentiated body cell.

📖 Regulations

A rule made and maintained by an authority.

The ethical issues of stem cell use and the regulation of their use

The use of embryonic stem cells raises **ethical issues**. These could include:

- regulations on the use of embryo stem cells;
- the use of **induced pluripotent stem cells;**
- the use of **nuclear transfer techniques.**

Many people believe an early embryo is classified as a human and therefore the removal of stem cells, which destroys the embryo, is equivalent to ending a human life.

Regulations ensure that the use of stem cells in research and therapy are carried out in accordance with the laws and rules in the UK. This helps to guarantee that all the procedures, safety and procurement of stem cells are carried out in the correct way. For example, embryos must not be allowed to develop beyond 14 days, around the time a blastocyst would be implanted in a uterus during in vitro fertilisation (IVF).

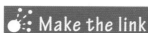 **Make the link**

Chapter 1.4 on gene expression explains in more detail about genes being 'switched on and off'.

Hint

Remember:
- the cells which can differentiate in plants and animals;
- specific examples of the therapeutic uses of embryonic and tissue stem cells;
- examples of ethical issues with the use of stem cells.

GO! Activity 1.5.1 Work individually to …

Restricted response

1. **Explain** the term 'differentiation'   1
2. **Name** the cells which are capable of differentiation in:
   a) a plant   1
   b) an animal   1
3. **Name one** part of the adult human body where stem cells are found.   1
4. **Give one** use of stem cells in research.   1
5. **Give two** therapeutic uses of stem cells.   2
6. **Describe one** way in which stem cell use is regulated.   1

Extended response

Give an account of ethical issues associated with the use of stem cells.   4

GO! Activity 1.5.2 Work in pairs to …

Research three therapeutic uses of stem cells. Once you have researched the uses, select one and find related data, using this to practice processing it. If appropriate, draw a conclusion on the success of the therapy and assess the reliability and relevance of the data.

GO! Activity 1.5.3 Work as a group to …

Debate the use of stem cells in research and therapy (remember to argue objectively from both sides of the debate). Make at least three arguments for and three against. Then split into two teams, and argue from either side. Try to predict the other team's arguments and write down counter-arguments.

After working on this chapter, I can:

1. Explain that cellular differentiation is the process by which a cell develops more specialised functions by expressing the gene's characteristic for that type of cell.

2. State that differentiation occurs in cells from meristems in plants and that stem cells in animals can differentiate.

3. Describe differentiation into specialised cells from meristems in plants and embryonic and tissue (adult) stem cells in animals.

4. State that stem cells are relatively unspecialised cells in animals that can continue to divide and can differentiate into specialised cells.

5. Describe the research into and therapeutic uses of stem cells by reference to the repair of damaged or diseased organs or tissues and state that stem cell research provides information on how cell processes such as cell growth, differentiation and gene regulation work.

6. State that stem cells can be used as model cells to study how diseases develop, or for drug testing.

7. State and discuss the ethical issues involving stem cell use and regulation, including use of embryos, induced pluripotent stem cells and nuclear transfer techniques.

1.6 The structure of the genome

Learning intentions

• Describe the components of the genome.
• Describe the function of coding sequences.
• Describe functions of some non-coding sequences.

Genomes

The **genome** of an organism is its hereditary information encoded in its DNA. DNA is packaged into chromosomes in the nuclei of eukaryotic cells. Modern techniques have allowed the DNA in the genomes of many different organisms, including humans, to be sequenced.

The human genome is about 3 billion bases long. Some of these base sequences are **genes**, which encode the information for synthesising proteins. However, most DNA sequences are non-coding and have functions related to the regulation of transcription and the formation of ribosomal and transfer RNA. Some non-coding regions have no known function.

📖 Genome

Hereditary information encoded in DNA.

📖 Genes

DNA sequences that code for proteins.

✷ Make the link

There is more about sequencing genomes in Chapter 1.8.

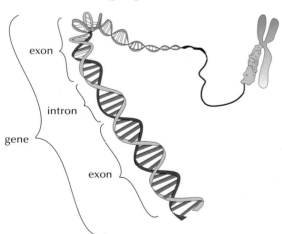

Figure 1.6.1 *A gene which contains exons and introns*

Genome structure

The structure of the genome is composed of both **coding** and **non-coding** regions. A summary is shown in **Figure 1.6.2**.

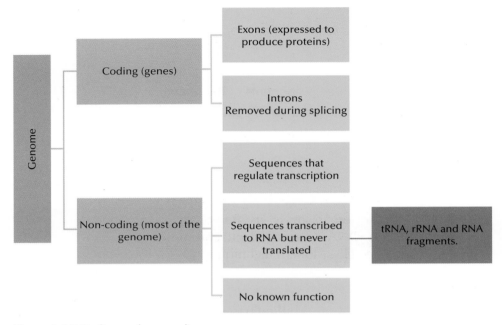

Figure 1.6.2 *Coding and non-coding sequences summary*

✳ Make the link

There is more about introns in Chapter 1.4.

Coding regions

The coding regions are the genes which contain sequences directly related to synthesising proteins. Notice that, within genes, there are non-coding regions called introns which may or may not be transcribed to mRNA. This system allows several proteins to be synthesised from the information within one gene.

📖 Coding sequences

Genes which code for proteins.

Non-coding regions

Most of an organism's DNA is non-coding and is involved in functions other than the encoding of protein.

1. Some sequences allow the production of molecules which can repress the transcription of genes, effectively turning these genes off.

2. Some encode the sequences needed to produce ribosomes or tRNA molecules.

3. Some have functions that are not currently fully understood.

📖 Non-coding sequences

Sequences of DNA which do not code for proteins.

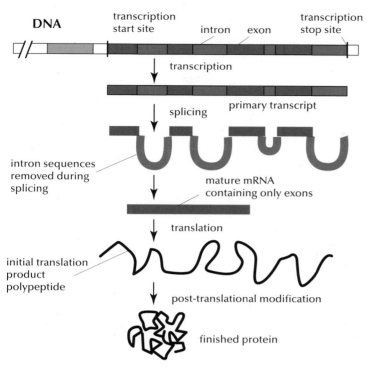

Figure 1.6.3 *Protein synthesis showing splicing of intron sequences*

GO! Activity 1.6.1 Work individually to ...

Restricted response

1. **State** the **two** types of sequences within the genome.   2
2. **Name** the molecules which are coded for by genes.   1
3. **Name two** forms of RNA which are not translated.   2
4. Describe the differences between an exon and an intron.   2

Extended response

Describe the structure of the genome.   5

GO! Activity 1.6.2 Work in pairs to ...

Create a coloured poster of the structure of the genome, highlighting the main components and explaining their functions to include coding and non-coding regions, introns and exons and the components of the non-coding region. Present this to the class, taking care to emphasise the differences in the regions and their functions.

After working on this chapter, I can:

1. Describe the genome of an organism as its hereditary information encoded in DNA.

2. State that DNA sequences that code for protein are genes.

3. Describe the structure of the genome as coding and non-coding sequences.

4. State that a genome is made up of genes and other DNA sequences that do not code for proteins.

5. State that within genes there are exons which are coding and introns which are non-coding and are removed from a primary transcript before translation.

6. Explain that non-coding sequences include those that regulate transcription and those that are transcribed to RNA but are never translated, such as tRNA, rRNA and RNA fragments.

7. State that some non-coding sequences have no known function.

1.7 Mutations

Mutations

Mutations are random changes in the genome that can result in no protein or an altered protein being expressed. These can range from a change in a single base to changes in chromosome structure or number. If the change to the genome results in an alteration to the organism's phenotype, it is called a mutant organism.

Mutation

A random and spontaneous change in the genome.

Mutagenic agent

Chemical or radiation, which increases the frequency of mutation in the genome.

Mutations are random, spontaneous and rare errors in genetic information. The frequency of the occurrence of these mutations can be increased by exposure to **mutagenic agents**, such as chemicals and radiation. Mutations can occur during the processes of DNA replication or gamete formation. Mutations cause alterations to the genome; this may contribute to the evolution of a species.

Single gene mutations

Substitution

A base replaced by another, with no other bases changing.

Insertion

An additional base inserted into the sequence.

Deletion

A base deleted from the sequence.

A single nucleotide base change results in single gene mutations. These involve the alteration of a DNA nucleotide sequence as a result of the **substitution**, **insertion** or **deletion** of nucleotides. The impact can be either minor or major, depending on the type of mutation. The table below shows single gene mutations and these are illustrated in **Figure 1.7.1**.

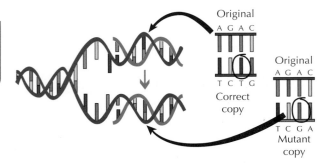

Figure 1.7.1 *A change in the genome*

Single gene mutation	DNA base sequence	Impact of gene mutation
None	ATG CGT CGA	
Substitution	ATG CCT CGA	Minor change to protein structure
Insertion	ATG CGG TCG	Major change to protein structure
Deletion	ATG CTC GA	Major change to protein structure

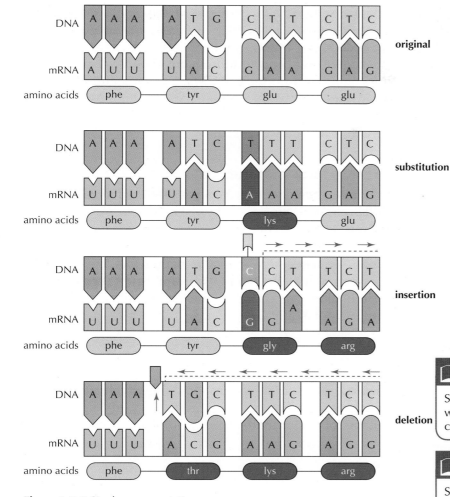

Figure 1.7.2 *Single gene mutations*

📖 **Missense**
Substitution of nucleotide which results in a changed codon.

📖 **Nonsense**
Substitution of nucleotide which results in a codon being changed to a stop codon.

📖 **Splice-site mutation**
Substitution of nucleotide at a splice site.

If a single nucleotide is substituted, then this will only change one codon, resulting in a minor change to the protein produced. Single-nucleotide substitutions include: **missense**, **nonsense** and **splice-site** mutations. Splice site mutations can alter post-transcriptional processing. The table shows single nucleotide substitutions.

Substitution	Change	End effect on protein produced
Missense	One codon to another	Different amino acid translated. Possible change in protein shape. However, may have no effect.
Nonsense	A codon to a stop codon	Shortens the protein. May become non-functional or its function will be changed.
Splice-site	Change in nucleotide at a splice site (between intron and exon)	May prevent splicing. This results in a very different protein being synthesised because introns may be left in the primary transcript.

📖 Frame-shift

Insertion or deletion of nucleotides which results in every subsequent codon to the right of the mutation in the base sequence being different and results in the synthesis of a different protein.

Nucleotide insertions or deletions result in **frame-shift** mutations or an expansion of a nucleotide sequence repeat. As a result, when the codon is translated at the ribosome into an amino acid, that amino acid and all subsequent ones are changed. This has an overall effect on the protein synthesised and may result in a faulty, non-functional or alternative protein. Mutations in some non-coding DNA sequences can result in changes to the way certain genes are expressed.

Nucleotide sequence repeats

Nucleotide sequence repeats are short DNA sequences that are repeated a number of times in a row. For example, a trinucleotide repeat is made up of 3-base-pair sequences, and a tetranucleotide repeat is made up of 4-base-pair sequences. An expansion of a nucleotide sequence repeat is a mutation that increases the number of times that the short DNA sequence is repeated. This type of mutation can cause the resulting protein to function differently or not at all.

Chromosome structure mutations

Chromosome structure mutations are those which affect whole chromosomes or sections of the genome. They are alterations to the structure of one or more chromosomes. They include:

📖 Duplication

Genes are copied and remain in the chromosome.

📖 Inversion

Chromosome breaks in two places and the segment is turned around before being rejoined.

📖 Translocation

Genes from one chromosome are added on to another chromosome.

- **Duplication**
- **Deletion**
- **Inversion**
- **Translocation**

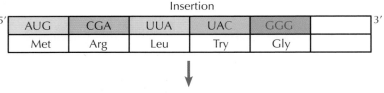

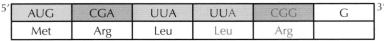

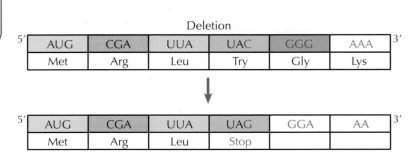

Figure 1.7.3 *Effects of frame-shift mutations on amino acid sequences*

Insertion and deletion are illustrated in **Figure 1.7.3**. Each of the mutations is described in the table below:

Mutation	Description
Duplication	Produced when extra copies of genes are generated on a chromosome.
Deletion	Results from the breakage of a chromosome in two places in which the genetic material becomes lost during cell division.
Inversion	A broken chromosome segment is reversed and inserted back into the chromosome.
Translocation	The piece of chromosome detaches from one chromosome and moves to a new position on another chromosome.

As the mutated chromosome segments are often large, many genes are often affected. The effect of this can be large alterations in the genome, and therefore the proteins produced, many of which will be incorrect or defective.

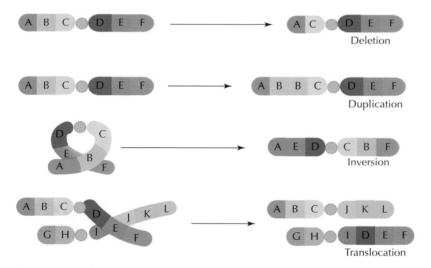

Figure 1.7.4 *Chromosome structure mutations*

The importance of mutations and gene duplication in evolution

Mutations cause random changes in the genome, the result of which is that they can sometimes cause new variations. This variation is the basis of evolution, as when mutations cause an advantageous change which allows an organism to be better adapted, it means those organisms survive and pass their genes on to the next generation.

Gene duplication is similarly important. Mutation in duplicated genes allows new genes to be produced while the original gene

📖 **Variation**

The differences in individuals within a species.

📖 **Polyploidy**

A condition in which the organism's cells have more than the diploid number of sets of chromosomes.

which was duplicated is retained. Therefore, mutations are a source of new **variation** and the basis of evolution.

Polyploidy

Polyploidy occurs when errors during the separation of chromosomes during cell division result in cells with whole genome duplications as shown in **Figure 1.7.5**.

Chromosome numbers

In sexual reproduction, gametes normally have a single set of chromosomes and can be described as haploid. Different species have different numbers of chromosomes in their haploid set. For example, there are seven chromosomes in the gametes of pea plants. The haploid number can be represented as **n**. When gametes undergo fertilisation, the resulting zygote is normally diploid **2n** because it has two sets of chromosomes.

Errors during the separation of chromosomes in cell division can result in whole genome duplications, as shown in **Figure 1.7.4**. In this example, gametes are produced with 2n chromosomes and they fertilise to produce a zygote with four sets of chromosomes (4n).

Organisms with sets of chromosomes above the diploid number are said to be polyploid.

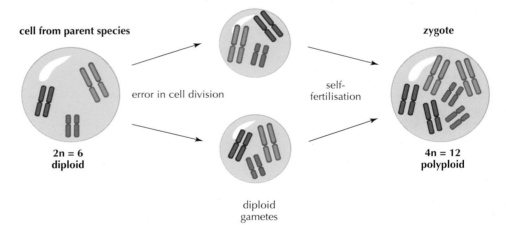

cell from parent species

error in cell division

self-fertilisation

zygote

2n = 6
diploid

diploid
gametes

4n = 12
polyploid

Figure 1.7.5 *Production of a polyploid*

Polyploidy is far more common in plants than in animals and has been particularly important in the origin of human food crop plants.

Artificial induction of polyploidy is a technique used to overcome the sterility of a hybrid species during plant breeding. The hybrid becomes fertile and can thus be further propagated, as shown in **Figure 1.7.6**.

The hybrid combines the qualities of its two parent species and polyploidy makes it fertile. Polyploidy examples include banana (triploid 3n) and potato (tetraploid 4n), as well as swede, oil seed rape, wheat and strawberry. The table below shows examples of polyploid crops.

Name of crop	Chromosome number	Type
Maize	20 (2 × 10)	Diploid
Wheat	42 (6 × 7)	Hexaploid
Rice	24 (2 × 12)	Diploid
Potatoes	48 (4 × 12)	Tetraploid
Bananas	33 (3 × 11)	Triploid
Sugar cane	80 (8 × 10)	Octoploid

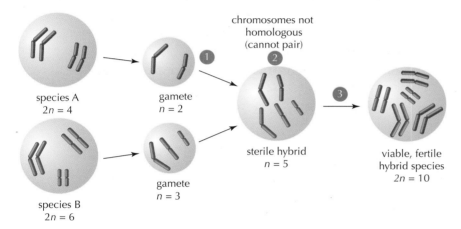

Figure 1.7.6 *Inducing polyploiidy to make a sterile hybrid fertile*

Importance of polyploidy in evolution and human food crops

As a result of having additional sets of chromosomes, organisms have an evolutionary advantage, due to their increased ability to mask harmful recessive alleles. In order to inherit a recessive trait, all of the alleles received must be recessive. In the case of polyploidy organisms, this will require the inheritance of more than two recessive alleles, reducing the odds of inheriting the characteristic or trait.

There are many examples of natural polyploidy in crops which have become important to us, both in terms of food production and economically. These polyploids tend to grow to be stronger, larger and more productive. Examples include strawberries, bananas and wheat.

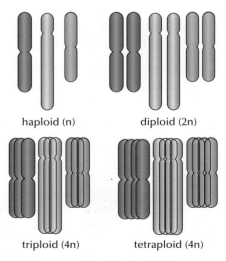

Figure 1.7.7 *Terms for the different numbers of sets of chromosomes*

(GO!) Activity 1.7.1 Work individually to …

Restricted response

1. Define the term mutation. 1

2. Look at the table below.

Mutation	DNA base sequence	Impact of change
None	ATG CGT CGA	
A	ATG CTC GA	1
B	ATG CGG TCG	2
C	ATG CCT CGA	3

    a. State the types of mutation A, B and C. 2
    b. State whether the impact is major or minor for 1, 2 and 3.

State whether the impact is major or minor for 1, 2 and 3. 2

3. **Explain** why single gene mutations are important in terms of evolution. 2

4. **Explain** how a frame-shift mutation affects the genome. 2

5. **Describe** the impact of nucleotide sequence repeat expansion upon amino acid sequence and protein production. 2

Extended response

1. **Describe** the **three** single gene mutation types. 3

2. **Name four** types of chromosome structure mutations and describe their effects. 6

3. **Explain** the term polyploidy and its importance in terms of evolution and human food crops. 5

(GO!) Activity 1.7.2 Work in pairs to …

1. **Research** and give a presentation on the uses for humans of polyploid food crops.

2. **Research** conditions caused by single gene and chromosome structure mutations.

3. **Research** reasons for geographical variation in incidence of post-weaning lactose intolerance or sickle-cell traits in humans as examples of point mutation.

(GO!) Activity 1.7.3 Work as a group to …

Create a coloured display poster of the different types of mutations. You must include the headings **single gene** and **chromosome** and diagrams to illustrate each, as well as the overall effect of the mutations on the organism.

After working on this chapter, I can:

1. State that mutations are random changes in the genome that can result in no protein or an altered protein being expressed. ◯ ◯ ◯

2. Explain how single gene mutations involve the alteration of a DNA nucleotide sequence as a result of the substitution, insertion or deletion of nucleotides. ◯ ◯ ◯

3. State that single-nucleotide substitutions include: missense, nonsense and splice-site mutations. ◯ ◯ ◯

4. Explain how nucleotide insertions or deletions result in frame-shift mutations and have major effects on protein. ◯ ◯ ◯

5. Explain that splice-site mutation can result in introns being left in mRNA. ◯ ◯ ◯

6. Name the chromosome structure mutations – duplication, deletion, inversion and translocation. ◯ ◯ ◯

7. Explain the importance of mutations and gene duplication in evolution in terms the importance of variation with a species. ◯ ◯ ◯

8. State that polyploidy occurs following errors during the separation of chromosomes during cell division and this can result in cells with whole genome duplications. ◯ ◯ ◯

9. Explain that polyploidy can allow organisms to avoid expression of harmful recessive alleles and so could be important in evolution. ◯ ◯ ◯

10. Explain that polyploid human crops are often bigger and and higher yielding. ◯ ◯ ◯

1.8 Evolution

Evolution

Evolution

Changes in organisms over generations as a result of genomic variations.

Evolution involves changes in organisms over generations as a result of genomic variations. This is the reason for the great diversity of organisms which have developed from simpler ancestors.

The evolution of a species happens as a continual accumulation of gradual changes. These eventually give rise to new species, which branch off from a common ancestor. Examples are shown below in **Figure 1.8.1**.

The modern horse evolved from a smaller ancestor which lived in tropical forests where its small size helped it to avoid predators. Its limbs were adapted to movement on moist ground. Its teeth were adapted for a soft diet which included fruit. As climate change reduced forest sizes, dry plains were formed, which provided new habitat opportunities. Ancestral horses that moved to the new habitats gradually evolved hooves and increased in their overall size to be able to run quickly on the hard ground to avoid predators. Their teeth became bigger and harder to allow the chewing of grass.

50 million years of evolution

ancestral horse
rainforest

modern horse
dry plains

Figure 1.8.1 *Evolution of the horse due to changes in climate*

Gene transfer

Vertical gene transfer, also called vertical inheritance, occurs when genes are transferred from parent to offspring as a result of sexual or asexual reproduction. This is the method through which genes are transferred in eukaryotes and prokaryotes.

In sexual reproduction, two parents who differ genetically produce offspring which will show more variation.

In asexual reproduction, a parent organism produces genetically identical copies of itself.

Some prokaryotes can exchange genetic material horizontally, resulting in rapid evolutionary change. This is called **horizontal gene transfer**. This method transfers genes from one cell to the next. (See **Figures 1.8.2**.)

📖 **Vertical gene transfer**

When genes are transferred from parent to offspring as a result of sexual or asexual reproduction.

📖 **Horizontal gene transfer**

Method used by prokaryotes to transfer genes from one cell to another.

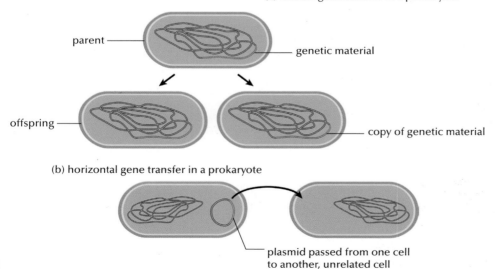

(a) vertical gene transfer in a prokaryote

parent — genetic material

offspring — copy of genetic material

(b) horizontal gene transfer in a prokaryote

plasmid passed from one cell to another, unrelated cell

Figure 1.8.2 *Vertical (a) and horizontal (b) transfer of genes*

Figure 1.8.2(b) shows how prokaryotes, such as bacteria, can exchange genetic information by passing plasmids from one cell to the next. Gene transfer is immediate and can result in rapid evolutionary change. The transfer of plasmids that carry resistance genes has led to widespread resistance to antibiotics in the bacterial population.

Prokaryotes and viruses can transfer sequences horizontally into the genomes of eukaryotes. For example, genetic engineering has led to the use of *Agrobacterium*, a bacterium well-known for its ability to transfer DNA between itself and plants. These inserted genes will then produce desirable characteristics within the plant species, such as increased nutrient content or resistances.

Selection

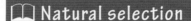

📖 **Natural selection**

Process where by organisms with advantageous genes (characteristics) are able to survive.

Natural selection is the non-random increase in the frequency of DNA sequences that increase chances of survival. Conversely, it could be said that it is also the non-random reduction in the frequency of deleterious sequences.

In other words, those organisms whose mutations are advantageous to survival will pass on their genes with the new characteristics. Organisms with non-advantageous mutations are less likely to survive and therefore less likely to pass on their genes.

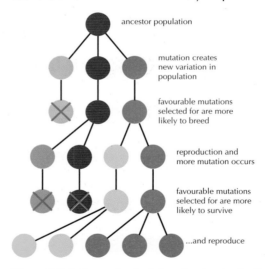

Figure 1.8.4 *General principles of natural selection*

These mutations in **Figure 1.8.4** are random. However, the selection of organisms which survive is non-random.

A term often associated with natural selection is survival of the fittest. This means members of a species with favourable characteristics that give them a selective advantage survive to breed.

Sexual selection is the non-random increase in frequency of DNA sequences that increase reproduction success.

Figure 1.8.5 *Peacocks display a large fan of brightly coloured feathers to attract a female*

As females produce eggs, which are much larger and less numerous than the male sperm, they have fewer opportunities to produce offspring. They also have to invest a lot of time and resources into the development and survival of their offspring and so it therefore makes sense for females to choose a sexual partner with genes linked to the most advantageous phenotypes. This will ensure favourable characteristics are passed on to the offspring, thereby increasing their chances of survival.

There are many examples of this with birds, such as brightly coloured feathers, being skilled in building a nest, or in dancing or singing. **Figure 1.8.5** shows how peacocks use colourful 'eyespots' in their tail feathers to attract a mate. The greater the number of 'eyespots', the more attractive the male is to the female (peahen).

Females choose a male that is the strongest within a group, due to its ability to protect its offspring and to produce young with similar characteristics.

> 📖 **Sexual selection**
>
> Process by which changes in genes (characteristics) increase chances of successful reproduction.

📖 Stabilising selection

A form of natural selection that favours the middle characteristics in variation.

📖 Directional selection

A form of natural selection that favours an extreme characteristic away from the middle.

📖 Disruptive selection

A form of natural selection that favours two extreme characteristics at the expense of the middle characteristics.

Figure 1.8.6 *Male kangaroos showing their strength to the females*

Selection pressures include factors such as predation or disease, which can remove organisms from a population. Depending on the selection type, there are different outcomes.

The differences in outcome as a result of **stabilising, directional** and **disruptive selection** are described in the table below.

Selection type	Description	Example	
Stabilising	A form of natural selection that favours the middle characteristics in a range of variation.	If a bird lays too few eggs, predators may eat them all. If a bird lays too many eggs, the young that hatch may not all be fed. So the number of average eggs laid remains at a middle value of the range.	
Directional	A form of natural selection that favours an extreme characteristic away from the middle of the range of variation.	At one time, white peppered moths were the most common colour phenotype, with dark-coloured moths in the minority. However, after the Industrial Revolution, when tree barks became darker due to the amount of black soot in the air, the white moths suffered an increase in predation while the dark moths were better camouflaged, therefore favouring an extreme, previously rare, characteristic.	

Disruptive	A form of natural selection that favours two extreme characteristics (at the expense of the middle characteristics) in a range of variation.	In a population of small mammals living at the beach where there is light-coloured sand interspersed with patches of tall grass, light-coloured mice that blend in with the sand would be favoured, as well as dark-coloured mice that can hide in the grass. Mixed-coloured mice, on the other hand, would not blend in with either the grass or the sand and, thus, would have an increased likelihood of being eaten by predators.	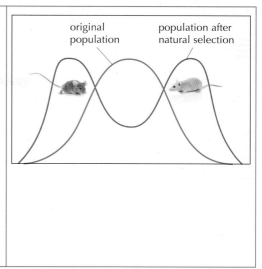

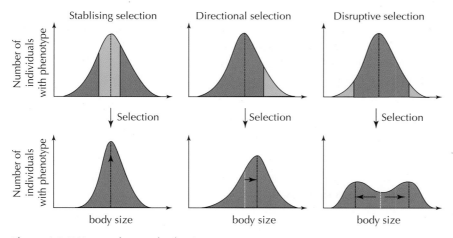

Figure 1.8.7 *Types of natural selection*

Genetic drift

The random increase and decrease in frequency of sequences, particularly in small populations, is a result of **neutral mutations** and **founder effects**. These can change the frequency of particular alleles within the population and therefore the number of organisms with a particular phenotype. These changes are called **genetic drift**.

Founder effects occur when groups become isolated or separated from the main or original group, and the splinter groups may have different frequencies of alleles, or leave behind some alleles altogether. The changed frequency of alleles in the new population make it different from the original group from which it separated.

> 📖 **Genetic drift**
>
> Random DNA sequence changes.

> 📖 **Neutral mutations**
>
> DNA sequence changes which are minor and therefore do not affect natural selection.

Figure 1.8.8 shows an example of a founder effect after a chance event has caused a reduction in population and allele frequency within that population. Therefore, in the new population the allele frequency has substantially changed compared with the original.

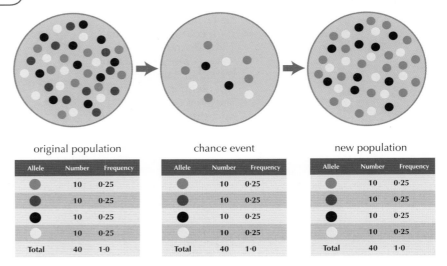

Figure 1.8.8 *Founder effect*

Figure 1.8.8 above shows two founder groups from a large original population. The phenotypic ratios in both groups are 2:1:1. However, the descendants of the splinter groups show a different phenotype ratio. This is due to the fact that fertilisation is a random process.

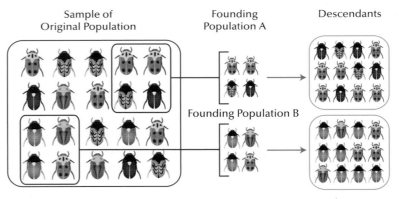

Figure 1.8.9 *Genetic drift in a beetle population which has two splinter groups*

Speciation

A species is a group of organisms which is capable of interbreeding and producing fertile offspring, and which does not normally breed with other groups.

For example, when a horse and a donkey breed together they produce an infertile mule as shown in **Figure 1.8.10**. Therefore a horse and a donkey, while closely related, are different species.

However, different dogs, as they are the same species, can produce fertile offspring as shown in **Figure 1.8.11**.

Figure 1.8.10 *Young infertile mule*

Figure 1.8.11 *Different dog breeds belong to the same species*

Speciation is the generation of new biological species by evolution as a result of isolation, mutation and natural selection.

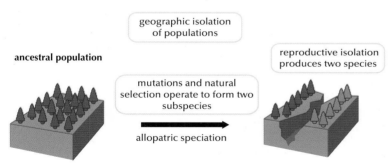

Figure 1.8.12 *Allopatric speciation*

Allopatric speciation occurs due to populations becoming physically separated. These separations are caused by geographical barriers. After the populations separate, the isolated populations could be exposed to a variety of selection pressures. They may also undergo genetic drift, as discussed earlier, or evolve new mutations, which may or may not be beneficial. If this new population has changed so much that it is unable to reproduce with the original population, it has become a new species.

Sympatric speciation differs from allopatric speciation in that geographical isolation does not take place. Instead, two species arise within the same habitat, but are separated by behavioural or ecological barriers.

> 📖 **Allopatric speciation**
>
> Evolutionary process where new species are formed due to a geographical barrier.

> 📖 **Sympatric speciation**
>
> Evolutionary process by which new species are formed due to an ecological or behavioural barrier.

Other examples of speciation

The London Underground mosquito (*Culex pipiens f. molestus*) is another interesting example of speciation. Over the last century, this mosquito has evolved from an overground mosquito (*C. pipiens*). Scientists believe that the original *C. pipiens* mosquitos moved into the London Underground and natural selection determined which variants of that population survived in the new environment. The adaptations selected included a loss of cold tolerance and a loss of the need to hibernate. The mosquitos also became adapted to the warmer conditions in the London Underground and started mating all year round. As well as this, the new species developed the ability to bite other organisms such as rats, humans and mice, unlike their ancestors, which only bit birds.

Culex pipiens f. molestus can be found in almost all of the underground systems in the world.

Hybrid zones

A **hybrid zone** is an area where two species frequently meet and may interbreed to produce fertile offspring. For example, **Figure 1.8.13** shows two species of crows, the carrion and the hooded, which are quite different in appearance and in niche. However, in hybrid zones as shown in **Figure 1.8.15**, they recognise each other as potential mates and may reproduce.

The offspring produced are generally less vigorous than their parents and so this selective disadvantage means they are unlikely to outcompete crows either within or outwith the hybrid zone. Hybrid zones are maintained by repeated hybridisation.

> 📖 **Hybrid zone**
>
> Overlapping geographical region where two populations frequently interbreed.

Figure 1.8.14 *Carrion and hooded crows*

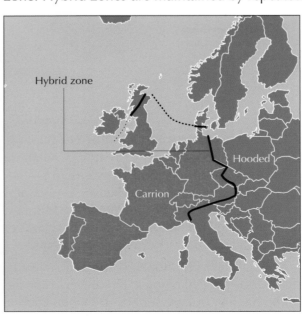

Figure 1.8.13 *Hybrid zones for two species of crow*

● Activity 1.8.1 Work individually to …

Restricted response

1. **Define** the term 'evolution'.   1
2. **State** the **two** general forms of gene transfer and describe the way in which they differ. 2
3. Describe natural selection in terms of the survival of the fittest.   3
4. Explain how sexual selection operates.   1
5. Define the terms 'genetic drift' and 'founder effect'.   2

Extended response

1. Give an account of the formation and maintenance of hybrid zones.   4
2. Give an account of speciation under the following headings:
   i. Allopatric speciation   3
   ii. Sympatric speciation   3

● Activity 1.8.2 Work in pairs to …

Research different definitions of the term 'species' (e.g. biological species concept, phylogenetic species concept) and the difficulty of applying species definition to asexually reproducing organisms.
Research the London Underground mosquito.

● Activity 1.8.3 Work as a group to …

Produce a presentation on chapter 1.8 under one of the following headings:

- evolution
- gene transfer
- selection
- genetic drift
- speciation
- hybrid zones

This may be in the form of a poster, PowerPoint, or any other suitable method. Create five slides for your presentation. Be sure to include keyword definitions as well as examples.

After working on this chapter, I can:

1. State that evolution is the gradual changes in organisms over generations as a result of natural selection.

2. Explain that gene transfer can be vertical (inheritance) from parent to offspring as a result of sexual or asexual reproduction; or horizontal (in prokaryotes), resulting in rapid evolutionary change.

3. State that prokaryotes and viruses can transfer sequences horizontally into the genomes of eukaryotes.

4. State that natural selection is the non-random increase in the frequency of DNA sequences that increases chances of survival for the next generation.

5. State that sexual selection is the non-random increase in the frequency of DNA sequences that increases reproductive success.

6. Describe the differences in outcome as a result of stabilising, directional and disruptive selection.

    a) Stabilising selection is a form of natural selection that favours the middle characteristics in a range.

    b) Directional selection is a form of natural selection that favours extreme characteristics away from the middle of a range.

    c) Disruptive selection is a form of natural selection that favours two characteristics at the expense of the middle characteristics in a range.

7. State that genetic drift is the random increase or decrease in frequency of sequences, particularly in small populations, as a result of neutral mutations and founder effects.

8. State that speciation is the generation of new biological species by evolution as a result of isolation, mutation and selection.

9. State that geographical barriers prevent gene flow in allopatric speciation.

10. State that behavioural or ecological barriers prevent gene flow in sympatric speciation.

11. Describe that hybrid zones form between two closely related species in which interbreeding occurs.

1.9 Genomic sequencing

You should already know:

- The genetic code of an organism is the sequence of nucleotide bases on its DNA.

Learning intentions

- Explain the term 'genomic sequencing'.
- Describe the importance of comparing sequence data, computer and statistical analyses (bioinformatics).
- Describe how evidence from phylogenetics and molecular clocks is used to determine the main sequence of events in evolution.
- State that the sequence of events can be determined using sequence data and fossil evidence.
- State that a comparison of sequences provides evidence of the three domains.
- State that comparison genomes from different organisms reveal that many genes are highly conserved across different organisms.
- Explain how personal genomics can be used in health care.
- Describe some of the difficulties with personalised medicine.

📖 Genomic sequencing

Ordering the sequence of nucleotide bases in a genome.

📖 Bioinformatics

Analysis of sequence data using computers and statistics.

📖 Phylogenetics

Study of evolutionary relatedness of species.

📖 Molecular clock

Graph which shows differences in sequence data for nucleic acids or proteins over time.

Genomic sequencing

Genomic sequencing is a process where the exact order of the nucleotide bases along an organism's genome is determined. The human genome project, finished in 2003, sequenced about three billion nucleotide bases and found that there are around 20,000 genes in the human genome.

The sequence of nucleotide bases can be determined for individual genes and for entire genomes. This is useful in terms of identifying the genes and genome sequences that are responsible for particular diseases and conditions; and to identify genomic mutations and variations.

To compare sequence data, computer and statistical analyses (**bioinformatics**) are required. This speeds up the process and enables whole genomes to be sequenced very quickly.

Phylogenetics and molecular clocks

All life on Earth is related to a common ancestor. Phylogenetics is the formal name for the field within biology that studies the patterns of relationships among organisms and reconstructs evolutionary history.

Evidence from **phylogenetics** and **molecular clocks** has been used to determine the main sequence of events in evolution.

Phylogenetic trees, such as in **Figure 1.9.1**, show the evolutionary relatedness of selected organisms in terms of time since the last common ancestor. Each branch in the tree represents the evolution of two distinct species from one common ancestor.

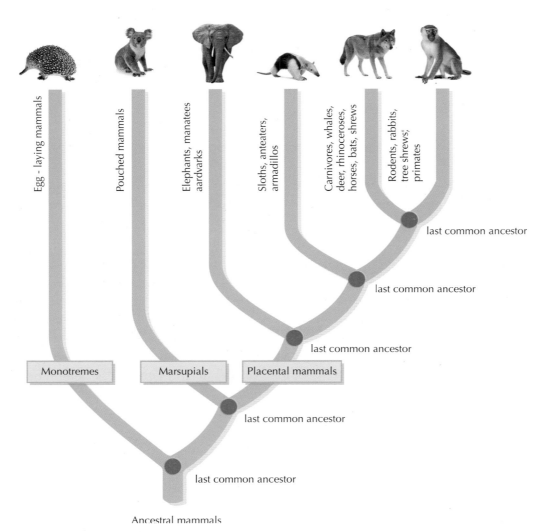

Figure 1.9.1 *A phylogenetic tree showing the evolution of some mammals*

Make the link

Sequence divergence is used to estimate the amount of time since lineages diverged. Studies have shown that the last common ancestor of the human and chimpanzee for example, lived between four and seven million years ago. In order to study this relatedness, highly conserved DNA sequences are used for comparisons of the distantly related genomes. The similarities of the human and chimpanzee chromosomes are shown in **Figure 1.9.6.**

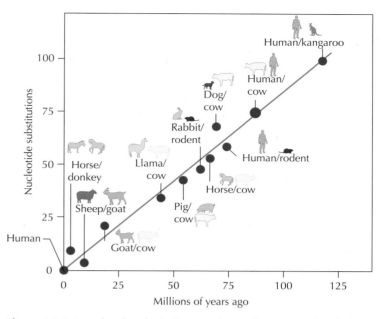

Figure 1.9.2 *A molecular clock showing the evolutionary relatedness of some organisms*

Over the course of millions of years, mutations may build up in any given stretch of DNA at a steady rate. If this mutation rate is reliable, change in the stretch of DNA could be used as a molecular clock. For example, **Figure 1.9.2** shows how organisms are related to each other by the number of nucleotide substitutions and the number of years since the pair of organisms had a common ancestor. The linear relationship between these two variables is displayed in the graph and can be used to determine evolutionary relatedness between two organisms.

The main sequence of events in evolution		
Event	**Millions of year ago (approximately)**	**Key details**
Last universal ancestor	3500	Last common ancestor evolves
Prokaryotes	3900–4500	Evolution of first prokaryotic cell
Photosynthesis	3000	First organism which could utilise the sun's energy to photosynthesise
Eukaryotes	1850	Appearance of cells with a true nucleus
Multicellular organisms	1200	Organisms with groups of interdependent cells first appear, followed by the evolution of animals, vertebrates and land plants

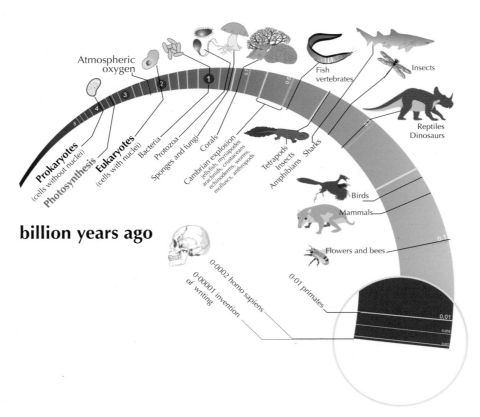

Figure 1.9.3 *Selected sequence of evolutionary events*

The sequence of events in the evolution of life can be determined using **sequence data** and fossil evidence.

Comparison of sequences provides evidence of the three **domains** of life (bacteria, archaea and eukaryotes).

Proposed in 1990 by Carl Woese, life on the planet Earth is classified into three different domains based on the differences in their genes. All three domains evolved from a common ancestor, as shown in **Figure 1.9.4**.

📖 **Sequence data**

Information about nucleotide base sequences and/or amino acids.

📖 **Domain**

A division of organisms based on shared similarities in DNA sequences.

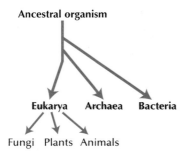

Figure 1.9.4 *Phylogenetic tree showing the three domains*

Genome comparison

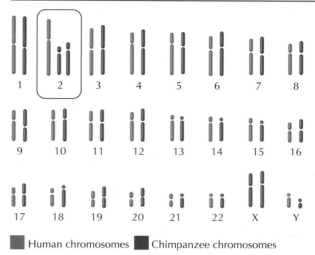

■ Human chromosomes ■ Chimpanzee chromosomes

Figure 1.9.4 *Comparison of human and chimpanzee chromosomes*

Figure 1.9.4 shows the human chromosome complement of 23 pairs, while chimpanzees, like many primates, have a complement of 24 pairs. Human chromosome 2 is a fusion of two chromosomes, 2a and 2b, that remained separate in the other primates.

Comparison of human and chimpanzee genomes reveals rapid change in genes for the immune system and regulation of neural development over the last six million years.

Comparison of genomes reveals that many genes are highly conserved across different organisms. Many organisms' genomes have now been sequenced and subsequent analysis and comparisons have shown similarities even in species which are outwardly very different. For example, the table below shows that the gene number in humans and mice is similar, as well as the approximate sizes of the genomes.

Comparison of genome sizes of different organisms			
Organism	Chromosome number	Estimated size (base pairs)	Estimated gene number
Human (Homo sapiens)	46	3 billion	~25,000
Mouse (Mus musculus)	40	2·9 billion	~25,000
Fruit fly (Drosophila thaliana)	8	165 million	13,000
Plant (Arabidopsis thaliana)	10	157 million	25,000
Roundworm (Caenorhabditis elegans)	12	97 million	19,000
Yeast (Sacchoromyces cerevisiae)	32	12 million	6,000
Bacteria (Escherichia coli)	1	4·6 million	3,200

Personal genomics and health

The human genome was sequenced in 2003. At that time, it cost £3 billion to sequence the genome. Due to advances in technology, an individual human genome can now be sequenced cheaply and quickly, for a fraction of that cost, and in days rather than years. Therefore, obtaining an individual's personal genome is relatively easy and this information could be used in a number of ways in the future.

One way in which this data could be used is to identify mutations within the genome. As discussed before, these can either be harmful (the changes result in wrong/no protein forming) or neutral (the change has no negative effect).

Analysis of an individual's genome may lead to **personalised medicine (pharmacogenetics)** through knowledge of the genetic component of risk of disease and the likelihood of success of a particular treatment. This means that identification of the genomic change responsible for a genetic disorder will enable a specific treatment to be applied, specific to that individual. Treatments will be personalised, specific and less likely to be ineffective.

However, many diseases and conditions arise from a combination of genetic and environmental factors, which can cause difficulties when it comes to treatment. Such diseases can also be complex and can be caused by other organisms, such as with viral infections.

> 📖 **Personalised medicine**
>
> Treatment which is based upon an individual's own genome.

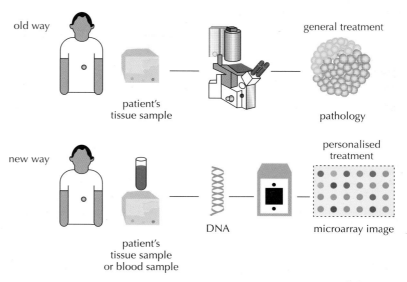

Figure 1.9.7 *Personalised diagnoses which could lead to personalised medicine*

Difficulties with personalised medicine

Difficulties with personalised medicine include distinguishing between neutral and harmful mutations in both genes and regulatory sequences, and in understanding the complex nature of many diseases.

There are also ethical issues to consider. One possible scenario is that genetic discrimination might take place by insurance companies and employers, which may refuse policies or work on the basis of potential conditions.

Finally, consideration must be given to the effect that knowledge of a particular potential genetic disease may have on subsequent quality of life and happiness.

⃝ Activity 1.9.1 Work individually to ...

Restricted response

1. Define 'genomic sequencing'.   1
2. State the term used for the statistical and computer-based treatment of sequence data.   1
3. Name the three domains of life.   3
4. Place the following sequence of events in order, starting with the earliest:

   multicellular organisms, photosynthesis, last universal ancestor, prokaryotes, eukaryotes   2

Extended response

5. Give an account of personalised genomics and medicine.   6
6. Give an account of phylogenetics and molecular clocks.   6

> **GO!** Activity 1.9.2 Work in pairs to ...

Research how sequencing technologies use techniques such as fluorescent tagging of nucleotides to identify the base sequence.

Research and create your own molecular clock or phylogenetic tree for a set of related organisms, using data from the internet. Ensure you select reliable data. This will be an important skill in your assignment.

> **GO!** Activity 1.9.3 Work as a group to ...

Debate the ethical issues surrounding the use of personal genomics in medicine and other areas.

After working on this chapter, I can:

1. State that genomic sequencing is the sequencing of nucleotide bases that can be determined for individual genes and entire genomes.

2. Describe the importance of comparing sequence data, computer and statistical analyses (bioinformatics) in terms of determining the order of evolutionary events.

3. State that evidence from phylogenetic trees and molecular clocks is used to determine the main sequence of events in evolution, using differences in nucleotide base substitutions.

4. State that the sequence of events of evolution can be determined using sequence data and fossil evidence.

5. State that a comparison of genomic sequences provides evidence of the three domains – bacteria, archaea and eukaryotes.

6. State that comparison genomes reveal that many genes are highly conserved across different organisms.

7. Explain that personal genomics can be used in health, in order to develop treatments that will be personalised, specific and less likely to be ineffective.

8. Describe difficulties with personalised medicine, such as distinguishing between neutral and harmful mutations in both genes and regulatory sequences, and in understanding the complex nature of many diseases.

2

Metabolism
and survival

2.1 Introduction to metabolic pathways

You should already know:

- Biological reactions are catalysed by enzymes.
- An enzyme is specific for its substrate.
- Enzymes have active sites complementary to their substrates.
- Enzymes can bring about the breakdown of substrates.
- Enzymes can bring about the synthesis of products.
- In the first stage of aerobic respiration, glucose is broken down to pyruvate.
- Cells are surrounded by membranes that regulate what can enter or leave a cell.
- Cell membranes consist of proteins and phospholipids.
- Diffusion occurs passively across cell membranes down a concentration gradient.
- Active transport uses energy for membrane proteins to move substances against a concentration gradient.

Learning intentions

- Describe the differences between an anabolic pathway and a catabolic pathway.
- State that enzyme-catalysed reactions can be reversible or irreversible.
- State that membranes form metabolic surfaces and compartments.
- Know that cell membranes have protein pores, protein pumps and enzymes embedded.

📖 Metabolism

All of the chemical reactions that take place within a living organism.

📖 Metabolic pathway

A chain of different biochemical reactions.

Anabolic and catabolic pathways

All the reactions which take place in a living cell are collectively called **metabolism**, which takes place via integrated and controlled enzyme-catalysed reactions. A series of these reactions is known as a **metabolic pathway**.

There are two basic types of metabolic pathway:

1. In **catabolic pathways**, a substrate is usually a large molecule and the products are small molecules. Energy is released. The following diagram shows a simple catabolic pathway:

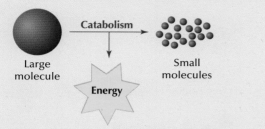

Figure 2.1.1 *A simple catabolic pathway*

The digestion of starch into glucose is a catabolic process, as shown below:

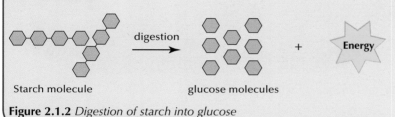

Starch molecule   glucose molecules

Figure 2.1.2 *Digestion of starch into glucose*

2. In **anabolic pathways**, small molecules are joined together to form larger molecules. Energy is required, as shown in the following diagram:

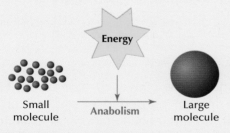

Figure 2.1.3 *A simple anabolic pathway*

The synthesis of proteins from amino acids is an anabolic process, as shown:

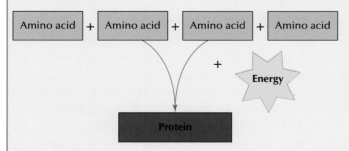

Figure 2.1.3a *Amino acids being joined together to synthesise protein*

Catabolic and anabolic pathways are often linked together in a cell to generate new substances required, as shown in **Figure 2.1.4**.

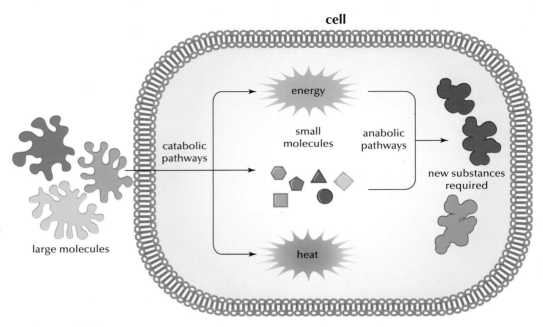

cell

Figure 2.1.4 *How catabolic and anabolic reactions can be linked*

Reversible and irreversible steps

Hint

Some of the energy released during catabolism is in the form of heat.

The reactions which make up a metabolic pathway are catalysed by enzymes. Sometimes these reactions can operate in both directions, because the enzyme can act on the product to produce the original substrate, as shown in **Figure 2.1.5**.

$$X \underset{}{\overset{E_1}{\rightleftharpoons}} Y \underset{}{\overset{E_2}{\rightleftharpoons}} Z$$

Figure 2.1.5 *Reversible enzyme controlled reactions*

In this example, chemical X is converted by enzyme E_1 into chemical Y, which is in turn converted into chemical Z by enzyme E_2. However, if the concentration of chemical Y becomes much lower than that of chemical X, enzyme E_1 can catalyse the conversion of Y to X. Similarly, enzyme E_2 can convert Y into Z, or Z into Y depending on their relative concentrations. This ensures that excess product is not formed unnecessarily, by using reversible reactions.

Glycolysis is the first stage in the break down of glucose in cells. In the process of **glycolysis**, there are ten different stages in the conversion of glucose to pyruvate. Three of these involve different reversible steps controlled by enzymes, whose actions depend on the relative concentrations of the substrates and products. These enzymes can catalyse their reactions both ways if required, so that the product can also serve as a substrate, which can fit into the enzyme's active site.

Glycolysis

The initial series of reactions in respiration which takes place in the cytoplasm and does not require oxygen.

In addition to the three reversible steps, there are also three irreversible steps. These are controlled by three different enzymes. One of these enzymes modifies glucose, after it diffuses into a cell across the membrane, trapping it inside the cell. The modified glucose cannot be converted back into glucose, and is constantly available for further cell metabolism.

These enzymes can act as 'regulators' in the process of glycolysis, controlling the rate and direction of enzyme-catalysed reactions.

Alternative routes

When glucose is plentiful, some of the stages of glycolysis which involve the regulatory, irreversible enzyme-catalysed steps can be 'bypassed'. The product formed from this bypass pathway eventually links up with the later stages of glycolysis as shown in **Figure 2.1.6**.

> **Hint**
> Enzyme regulation is very common in living cells. Look for examples elsewhere in the course.

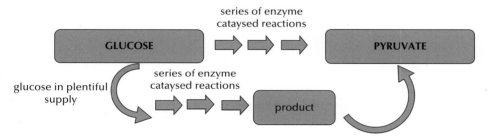

Figure 2.1.6 *Bypass pathway when glucose is in plentiful supply*

Cell membranes

Membranes form surfaces separating what is outside a cell from its internal, living contents. They are also highly selective about what can enter and leave a cell, and so therefore have a regulatory function.

In addition, membranes allow areas of the cell to be isolated into small compartments, where metabolic pathways can take place. These compartments allow metabolic pathways to operate rapidly, allowing high concentrations of substrates and products.

Cell **organelles** may be surrounded by double membranes, as shown by the examples in **Figure 2.1.7**.

> **Hint**
> Cells in the brain often use this kind of bypass to make sure the high demand for energy is satisfied.

> **Hint**
> Compartmentalisation allows for different local environments to be formed within a cell so that different, sometimes incompatible, pathways can go on at the same time.

> **Organelle**
> Subcellular component within cell cytoplasm which carries out a specific role.

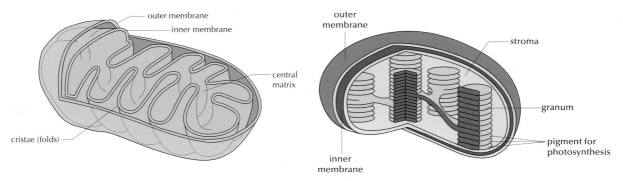

Figure 2.1.7 *Mitochondrion (left) and chloroplast (right) with double membrane layer surrounding their internal compartments*

📖 Citric acid cycle

A cyclical series of reactions operating under aerobic conditions which occur in the matrix of the mitochondrion.

📖 Calvin cycle

A series of linked reactions which form part of the reactions of photosynthesis. Carbon dioxide is 'fixed'.

Make the link

It is now thought organelles like these may originally have been independent cells.

Make the link

These organelles feature in later sections on respiration and photosynthesis.

🔍 Hint

Increasing the surface area of a cell, tissue or organ has a dramatic effect on the SA/V ratio. Think of the human lungs or the leaf of a plant.

The compartmentalisation of a mitochondrion allows the metabolic pathways associated with **aerobic respiration** to take place in isolation. In the same way, a chloroplast allows the reactions of photosynthesis to take place essentially outwith the cytoplasm of the plant cell.

When membranes are highly folded within small compartments, the surface area available for metabolic pathways is hugely increased. **Figure 2.1.8** shows the effect of changing the surface area, without changing the volume of an model cell on the ratio of the surface area/volume ratio (SA/V). Notice that an increase in the surface area while keeping the volume constant has a massive effect on the SA/V.

surface area increases while total volume remains constant

Total surface area (height x width x number of sides x number of boxes)	150	750
Total volume (height x width x length x number of boxes)	125	125
Surface-to-volume ratio (surface area / volume)	1.2	6

Figure 2.1.8 *The effect of increasing compartments on the ratio of surface area to volume ratio*

Membrane structure

The cell membrane is composed of two layers of **phospholipid molecules** within which are protein molecules arranged in different ways. Phospholipids are important for many of the cell membrane's properties. These molecules are constantly in a state of motion, and proteins of varying size and function are randomly found within and on them. **Figure 2.1.9** shows the current view of this fluid-mosaic model of the membrane structure:

> 📖 **Phospholipid**
>
> A molecule of fat which has a phosphate group attached.

> 🔍 **Hint**
>
> Remember that no ATP is used up in the process of diffusion, which is passive and does not need additional energy.

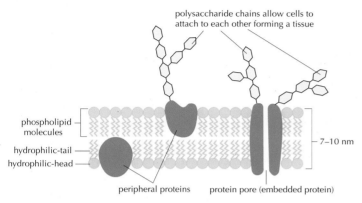

Figure 2.1.9 *Fluid mosaic model of cell membrane*

Many substances, such as oxygen and carbon dioxide, can pass across the cell membrane easily by simple diffusion, moving down a concentration gradient. Such substances can travel through the phospholipid layers as shown in **Figure 2.1.10**, opposite.

Protein pores

Within the cell membrane are **channel proteins**, which have pores in them that allow larger molecules to pass from an area of high concentration to an area of low concentration. This is shown in **Figure 2.1.11**. These larger molecules such as glucose, are unable to diffuse through the phospholipid layers.

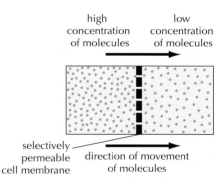

Figure 2.1.10 *Simple diffusion across the phospholipid bilayer*

> 📖 **Channel proteins**
>
> Proteins which span the phospholipid layers and allow large molecules to pass through their pores.

> 🔍 **Hint**
>
> Channel proteins are specific for the molecules they will allow to pass through their pores.

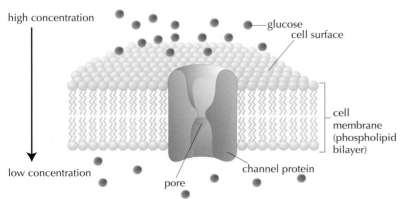

Figure 2.1.11 *Membrane with protein pore showing glucose molecules diffusing through the pore*

Protein pumps

Sometimes a cell moves molecules against a concentration gradient using **active transport**. This process requires energy to cause a change in the shape of the appropriate **protein pump**, so that it forces the movement of the molecule across the cell membrane against the concentration gradient. This is shown in **Figure 2.1.12**. Protein pumps are also specific for the molecules and ions they transport, such as potassium and sodium, allowing chemical and electrical gradients to develop.

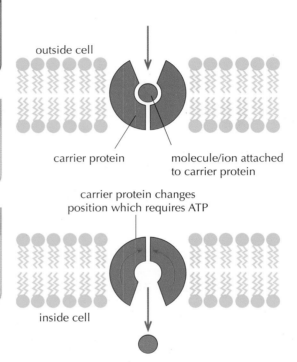

Figure 2.1.12 *Active transport using a protein pump*

Enzymes in cell membranes

Within cell membranes, enzymes are present which perform the same type of function as elsewhere in a cell: converting one molecule into another. A good example of such an enzyme is **ATP synthase**, found in all organisms which use aerobic respiration. It is also important in the process of photosynthesis. This enzyme may be located, for example, on the membranes of bacteria, inner membranes of mitochondria and inner membranes of chloroplasts. **Figure 2.1.13** shows the enzyme ATP synthase in a membrane.

📖 **Active transport**

An energy-demanding process in cells which moves molecules against a concentration gradient.

📖 **Protein pump**

Protein embedded in the phospholipid layers which can transport molecules across the cell membrane.

🔍 **Hint**

The energy to drive active transport comes from the breakdown of ATP.

🔍 **Hint**

Minerals, some sugars, and most amino acids can be moved into a cell by active transport.

📖 **ATP synthase**

An enzyme which can catalyse the synthesis of ATP from ADP.

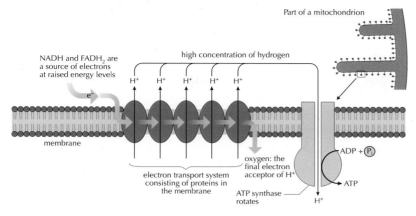

Figure 2.1.13 *ATP synthase in the cell membrane*

> ### Make the link
> You will study ATP synthase in more detail in Section 2.2 on respiration.

Activity 2.1.1 Work individually to:

Restricted response

1. State the difference between an anabolic pathway and a catabolic pathway. **2**
2. State **one** advantage of an enzyme catalysing a reversible reaction. **1**
3. Describe how an enzyme whose action is reversible can regulate a (metabolic) pathway. **1**
4. Explain how 'compartmentalisation', as it applies to the cell membrane may be useful in cell metabolism. **1**
5. At the start of the experiment the concentration of X and Y inside the cells was 40 millimoles per litre (mmol/L). The table below shows some data for the rate of uptake, (measured in millimoles per hour [mmol/h]) of two substances, X and Y, across a cell membrane. For each experiment, the cells were placed in a known concentration of either X or Y. The different rates of uptake are shown.

Concentration outside cell mmol/L	20		40		60		80		100		120	
Substance	X	Y	X	Y	X	Y	X	Y	X	Y	X	Y
Rate of uptake mmol/h	0	20	0	40	10	60	20	70	25	75	40	80

   a) On suitable graph paper plot this data as a line graph. **4**
   b) Under what condition will substance X be taken into the cell? **1**
   c) Name the process by which substance X is taken into the cell. **1**
   d) Name the process by which substance Y is taken into the cell and give evidence which supports your answer.. **2**

Extended response

Describe the arrangements and functions of molecular components of membranes in cells. **6**

GO! Activity 2.1.2 Work in pairs to:

Produce a set of flash cards which can be used for revision purposes. Each pair should chose a different section of the chapter to ensure that all the content is covered. The website: http://quizlet.com/ is excellent for this purpose.

GO! Activity 2.1.3 Work in groups to

Produce a poster to show the different ways in which substances can move across the cell membrane.

After working on this chapter, I can:

1. Define metabolism as the integrated chemical reactions within cells.

2. State that a metabolic pathway is a sequence of enzyme-controlled reactions in cell metabolism.

3. Describe anabolic pathways as energy requiring.

4. Describe catabolic pathways as energy releasing.

5. State that some metabolic pathways have reversible steps.

6. Explain that reversible steps can regulate the concentration of substances in a metabolic pathway.

7. State that some metabolic pathways have irreversible steps.

8. Explain that some irreversible steps can help to regulate metabolism by limiting the direction a pathway can take.

9. State that some metabolic pathways have alternative routes.

10. Describe the structure of cell membranes in terms of phospholipids and proteins.

11. State that membrane proteins include pores, pumps and enzymes.

12. Explain that some molecules move through the pores in membranes.

13. State that pumps carry molecules across membranes by active transport.

14. Give the example of ATP synthase as a membrane-bound enzyme.

15. State that membranes form metabolic surfaces and compartments in cells.

16. Explain the advantage of compartments in terms of the increase in surface area they can provide.

17. State that compartments allow for high concentrations of metabolic substances and increased rates of metabolic reactions.

2.2 Control of metabolic pathways

You should already know:

- An enzyme is specific for the substrate on which it acts.
- Enzymes are biological catalysts made by all living cells.
- Enzymes speed up cellular reactions or are unchanged by the process.
- The active site of an enzyme is complementary to its specific substrate.
- Enzymes are involved in degradation and synthesis reactions.
- Enzyme activity is affected by temperature, pH, and substrate concentration.

Learning intentions

- State that metabolic pathways can be controlled.
- Understand the different ways in which control of metabolic pathways may be achieved.
- Describe the action of an enzyme in terms of induced fit.
- State the role of the active site of an enzyme.
- State that the active site has a low affinity for the product(s) but a high affinity for the substrate(s).
- State that the activation energy of an enzyme-catalysed reaction is lowered by the enzyme.
- Explain the principle of feedback inhibition.
- State that enzymes may act in groups.
- State that enzymes may act as multi-enzyme complexes.
- Describe the different ways in which inhibitors can regulate enzyme activity.

Control of metabolic pathways

Since metabolic pathways are a series of enzyme-catalysed reactions, it is possible to control these reactions by regulating the activity of the enzymes involved. If the enzyme is present, the reaction proceeds but if the enzyme is absent, the reaction stops.

Like all proteins, an enzyme is coded for by a gene which may be 'switched off' or 'switched on' depending on whether or not the enzyme is required. For example, the gene which codes for an enzyme involved in digestion may not be switched on if the substrate is absent since it is not required. The vital process of cellular respiration goes on all the time and so the genes associated with the many enzymes of cellular respiration are constantly switched on.

Enzyme action

Enzymes only act on one specific substrate which is complementary to and fits into the active site. This action is called an **induced fit** and makes use of the fact that the active site is not a rigid arrangement but can change shape as shown in **Figure 2.2.1**.

> ### 📖 Induced fit
> Describes the change in shape of the active site when a specific substrate fits.

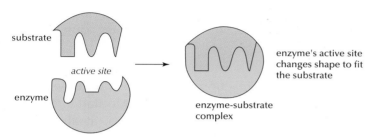

Figure 2.2.1 *Induced fit of enzyme's active site interacting with specific substrate*

In order for two or more substrates to fit into the active site, they have to be positioned in such a way that they can both fit into the active site and be acted upon by the enzyme. This is called **orientation** and makes sure the reactants are in the best possible position for the reaction to proceed as shown in **Figure 2.2.2**.

> ### 📖 Orientation
> The position in space of the reactants in an enzyme-catalysed reaction to ensure the reaction proceeds.

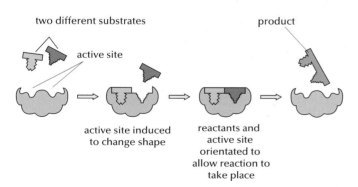

Figure 2.2.2 *Induced fit with two different substrates. Active site has high affinity for substrates but low affinity for product*

Activation energy

In order to make some reactions take place, a small input of energy is required to activate the process. When the reactants have reached a threshold energy level, at the **transition state** as shown in **Figure 2.2.3**, the reaction can proceed and products are formed.

> ### 📖 Transition state
> The highest energy stage in a reaction when the bonds in the reactants are breaking.

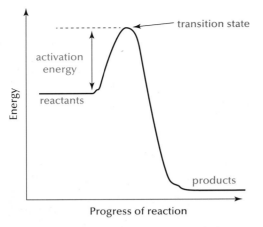

Figure 2.2.3 *Activation energy is needed to start reaction*

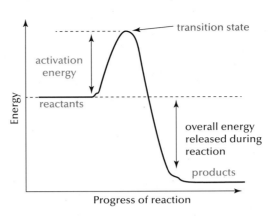

Figure 2.2.4 *An enzyme lowers the activation energy*

In an enzyme-catalysed reaction, the presence of the enzyme lowers this activation energy, making it easier to start a reaction which otherwise might proceed slowly or not at all. This is shown in **Figure 2.2.4.** Notice the enzyme does not affect the initial or final energy values for the reactants or products, nor does it affect the overall energy released.

Factors affecting enzyme action

The direction and rate of an enzyme-catalysed reaction are influenced by both the concentration of the substrate and the concentration of the product(s). **Figure 2.2.5** shows the effect of increasing substrate concentration on the rate of an enzyme-catalysed reaction when other variables are kept constant.

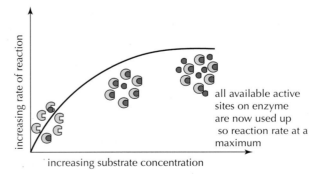

Figure 2.2.5 *Effect of increasing substrate concentration on the rate of reaction*

> ### 🔍 Hint
> Remember that other factors, such as pH and temperature, can affect enzyme activity.

As the substrate concentration increases, more and more of the enzyme's active sites become used, until a point is reached when no more are available. At this point, some other factor needs to be altered to make the reaction go any faster.

Another way an enzyme-catalysed reaction can be regulated occurs when the product of the last reaction in a metabolic pathway inhibits the enzyme which catalyses the first reaction of the pathway. This is called **end-product inhibition**, and ensures that levels of products meet the supply requirements of the cell without excess being produced, as shown in **Figure 2.2.6**.

🔍 **Hint**

The concept of factors such as concentration and temperature limiting reaction rates occurs throughout biological systems.

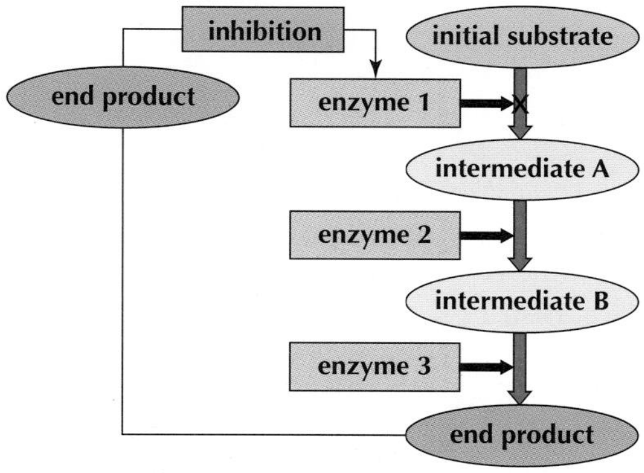

Figure 2.2.6 *End-product inhibition of a metabolic pathway*

📖 **End-product inhibition**

A mechanism used by a cell in which an enzyme that catalyses the conversion of a substrate into a product becomes inhibited when that product accumulates.

The product of the last reaction binds to a site other than the active site of the enzyme, which catalyses the first reaction changing the shape of the active site as shown in **Figure 2.2.7**.

This means the attachment of the first substrate to the first enzyme is now less likely and the reaction slows down or stops altogether. However, when the supply of the final product falls below a threshold level, its attachment to the first enzyme no longer occurs and the active site can once again bind with its normal substrate, to allow the reaction to proceed.

Make the link

End-product inhibition is a form of negative feedback. You will come across this often, for example in section 2.? on maintaining metabolism.

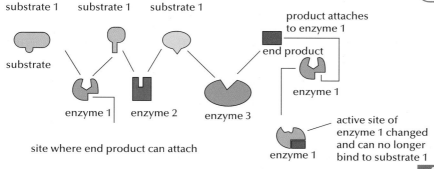

Figure 2.2.7 *Action of end product causing change in shape of the active site of enzyme 1*

The obvious advantage of using end product inhibition is to allow control of a metabolic pathway. Since many enzyme reactions are reversible this mechanism is one way of driving a reaction, as well as controlling the rate of the reactions.

Make the link

See Chapter 2.4, part 2, which deals with the chemical reactions of aerobic respiration in which this multi-enzyme complex operates.

It is possible for enzymes to work in groups called complexes. A multi-enzyme complex is a distinct and highly ordered set of enzymes, which may catalyse successive steps in a metabolic pathway. The conversion of pyruvate to acetyl coenzyme A involves a multi-enzyme complex called the pyruvate dehydrogenase complex, as shown in **Figure 2.2.8**.

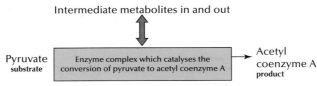

Intermediate metabolites in and out

Pyruvate **substrate** — Enzyme complex which catalyses the conversion of pyruvate to acetyl coenzyme A → Acetyl coenzyme A **product**

Figure 2.2.8 *Pyruvate dehydrogenase multi-enzyme complex*

Multi-enzyme complexes allow different enzymes to form 'clusters' in particular parts of a cell, and also for intermediate metabolites to be channelled from the active site of one enzyme to the active site of the next enzyme quickly and efficiently.

Control of metabolic pathways through inhibitors

The activity of many enzymes can also be controlled by the binding of specific molecules called **inhibitors**. Inhibition of an enzyme is a way of controlling biological systems, by affecting the enzyme-catalysed reactions associated with them. There are two important types of inhibitors.

Competitive inhibitors

Molecules which compete with the normal substrate for the active site of an enzyme, but which don't permanently disable the enzyme are called **competitive inhibitors**. Competitive inhibitors are very similar in their shape to the normal substrate of a particular enzyme, and will compete with these substrate molecules to occupy the active site as shown in **Figure 2.2.9**. The enzyme either binds to the normal substrate or the competitive inhibitor, but not both.

Since a competitive inhibitor is competing with the normal substrate for the active site, increasing the substrate concentration in the presence of an inhibitor will cause the rate of enzyme activity to increase again.

Non-competitive inhibitors

Some enzymes have two sites: one for the normal substrate and another for an inhibitor. The inhibitor can alter the shape of the active site so that the enzyme will no longer function, as shown in **Figure 2.2.10**. Such inhibitors are termed **non-competitive**. A non-competitive inhibitor binds to a site somewhere other than the active site so that the enzyme can no longer bind to the normal substrate.

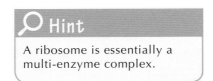

Hint

A ribosome is essentially a multi-enzyme complex.

Inhibitor

General term for a substance which can slow down or completely halt the activity of an enzyme.

Hint

Many useful drugs are inhibitors.

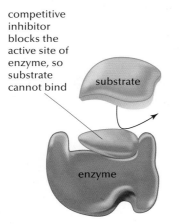

competitive inhibitor blocks the active site of enzyme, so substrate cannot bind

substrate

enzyme

Figure 2.2.9 *Action of a competitor inhibitor*

Competitive inhibitor

Substance which binds reversibly to the active site of an enzyme, thus reducing the quantity of active enzyme available.

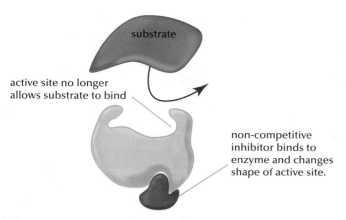

active site no longer allows substrate to bind

substrate

non-competitive inhibitor binds to enzyme and changes shape of active site.

> 📖 **Non-competitive inhibitor**
>
> Substance which binds irreversibly to an area other than the active site of an enzyme, causing a change in the shape of the active site so that the normal substrate no longer fits.

Figure 2.2.10 *Action of a non-competitor inhibitor.*

Since the inhibitor in this case alters the shape of the active site, increasing the substrate concentration will not increase the rate of the enzyme-catalysed reaction. This type of inhibition is irreversible.

GO! ## Activity 2.2.1 Work individually to:

Restricted response

1. Give **two** reasons why it is advantageous for the stomach enzyme pepsin to be synthesised only when protein is present in the food eaten. **2**

2. Describe the effect an enzyme has on the activation energy of an enzyme-catalysed reaction. **1**

3. The following graph shows the activity of an enzyme-catalysed reaction, as the substrate increases in the absence of any inhibitor and in the presence of inhibitors. **3**

Graph — y-axis: Rate of reaction; x-axis: Substrate concentration. Curves labelled A, B and C.

Using the letters, identify each of the reactions as being without inhibitor, with a competitive inhibitor and with a non-competitive inhibitor and give a reason for your answer to each.

4. The following graph shows the total mass of product produced in an enzyme-catalysed reaction at two different temperatures over a period of 10 minutes.

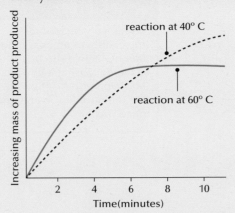

a) Describe how the mass of product changes over the 10-minute period at each temperature. **4**

b) Explain how the difference in the total amount of product produced after 10 minutes has arisen. **2**

Extended response

Give a description of the induced fit model of enzyme action. (Labelled diagrams may be used where appropriate.) **6**

GO! Activity 2.2.2 Work in pairs to:

Generate flow-charts to explain the sequence of events involved in some of the processes discussed in this chapter, such as end-product inhibition, multi-enzyme activity and different types of inhibitor action.

GO! Activity 2.2.3 Work in groups to:

Produce a set of 'bingo cards' for class use by co-ordinating groups to cover different topics. Each group should produce a set of bingo cards for their own topic. Each group should also produce a set of 'clues' for all the terms on the cards, to be used by the 'caller'. The caller will randomly read out the clues for the players to link to the terms on their cards. The caller might be your teacher or a fellow-student. The grids can be laminated for easy re-use. A good source of templates can be found at http://www.bingocardcreator.com/bingo-cards/biology

After working on this chapter, I can:

1. State that metabolic pathways are a series of enzyme-catalysed reactions.

2. State that one way to control metabolic pathways is to regulate enzyme activity.

3. Describe how enzyme production may be switched on/off depending on immediate requirements.

4. Explain the induced fit model of enzyme action.

5. Understand what is meant by the orientation of the reactants in an enzyme-catalysed reaction.

6. Draw a graphical representation of how an enzyme facilitates a reaction by lowering the activation energy.

7. State that enzymes do not change the initial and final energy of the reactants.

8. State that enzymes do not change the overall energy released during a reaction.

9. State that the direction of an enzyme-catalysed reaction is influenced by both the concentration of the substrate and the concentration of the product.

10. State that the rate of an enzyme-catalysed reaction is influenced by both the concentration of the substrate and the concentration of the product.

11. Illustrate, using a flow-chart, how end-product inhibition of a metabolic pathway may take place.

12. Explain how an enzyme's active site might be changed by the final product of the reaction it is catalysing.

13. Understand the concept of a multi-enzyme complex.

14. Explain how an inhibitor can control enzyme activity.

15. Explain the difference between a competitive and non-competitive inhibitor.

2.3 Cell respiration 1

You should already know:

- Respiration is an enzyme-controlled process which occurs in living cells.
- Respiration releases energy from glucose.
- Aerobic respiration takes place in the presence of oxygen.
- In the absence of oxygen, fermentation takes place.
- Adenosine triphosphate (ATP) is a universal energy carrier molecule found in living cells.
- Energy released from glucose is used to generate ATP from ADP and inorganic phosphate (Pi).
- The chemical energy stored in ATP can be released by breaking it down to ADP and Pi.
- Energy released from ATP can be used for cellular activities including muscle contraction, cell division and protein synthesis.
- ATP can be regenerated during respiration.

Learning intentions

- State that glucose is the main respiratory substrate in cell respiration.
- State that glucose is broken down during cell respiration.
- State that as glucose is broken down, hydrogen ions and electrons are removed by dehydrogenase enzymes.
- State that the hydrogen ions removed by the dehydrogenase enzymes are passed to a carrier molecule, (such as NAD or FAD.)
- State that adenosine triphosphate (ATP) is synthesised during the breakdown of glucose.
- State that the pathways of cell respiration are present in all three domains of life.
- Explain that the pathways of cell respiration provide energy and are connected to many other pathways.
- Describe the role of ATP in the transfer of energy.
- State that ATP can phosphorylate other molecules.
- Describe the working of the membrane enzyme ATP synthase.
- State that the final electron acceptor is oxygen, which combines with hydrogen ions and electrons to form water.

Cell respiration

Cell respiration is a vital process which occurs in living cells from all three domains: bacteria, archaea and eukaryotes. It consists of a series of reactions making up a metabolic pathway responsible for generating a source of energy. The main substance which is broken down to release energy is glucose, the **substrate**. A simple word equation for cellular respiration is shown below:

> ### 📖 Substrate
> Molecule on which an enzyme acts.

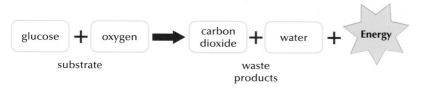

Figure 2.3.1 *Simple word equation for cell respiration*

When glucose is broken down to carbon dioxide and water, with oxygen present, adenosine triphosphate (ATP) is synthesised, as hydrogen ions and electrons are removed using **dehydrogenase** enzymes. The hydrogen ions and electrons are picked up by a carrier molecule called **nicotinamide adenine dinucleotide (NAD)** or **flavine adenine dinucleotide (FAD)**.

> ### 📖 Dehydrogenase
> Enzyme which catalyses the removal of hydrogen from a substrate.

The energy released during cell respiration is used to drive activities such as protein synthesis, active transport, mitosis, muscle cell contraction, maintenance of body temperature and growth.

Role of ATP

ATP synthesised during cell respiration transfers energy immediately to where it is needed. As it makes this transfer, the ATP is converted to adenosine diphosphate (ADP), and a molecule of inorganic phosphate (P_i) as shown:

> ### 📖 Nicotinamide adenine dinucleotide (NAD)
> Carrier molecule which accepts hydrogen ions.

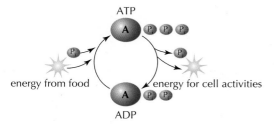

Figure 2.3.2 *Transfer of energy from ATP breakdown*

> ### 📖 Flavine adenine dinucleotide (FAD)
> Carrier molecule which accepts hydrogen ions.

As well as energising other processes, ATP can also add a phosphate group to a molecule. This is called **phosphorylation**, and is shown in **Figure 2.3.3**. Phosphorylation increases the energy content of the molecule to which the phosphate group is added.

> ### 📖 Phosphorylation
> Addition of a phosphate group to a molecule.

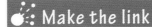

Hint

Remember that enzymes are specific for the substrate on which they act.

Make the link

Expect to see the ATP–ADP cycle anywhere in a cell process where energy transfer is involved.

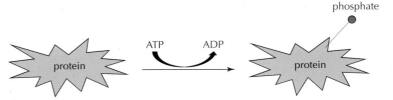

Figure 2.3.3 *Example of phosphorylation by ATP of a protein molecule.*

ATP is not a very stable molecule, and is used up as it is produced. In addition, the storage of ATP inside a cell sets up a concentration gradient, drawing water into a cell by osmosis. Cells are unable to store large quantities, and rely on the constant regeneration of ATP from ADP and P_i using the energy derived from cell respiration.

ATP synthase

The generation of ATP is brought about by the action of an enzyme called **ATP synthase**. It is large enough to be visible through an electron microscrope. This enzyme sits in the inner membrane of the mitochondrion, catalysing the synthesis of ATP from ADP and P_i. The energy to drive the process is obtained by a flow of hydrogen ions across the membrane, through a channel in the ATP synthase and down a concentration gradient as shown opposite.

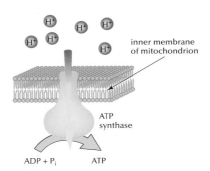

Figure 2.3.4 *Synthesis of ATP using enzyme ATP synthase*

ATP synthase

Membrane-bound enzyme which catalyses the synthesis of ATP from ADP and inorganic phosphate.

Hint

The hydrogen ions come from the metabolic pathway of cell respiration, and will be discussed later in this unit.

Hint

To maintain the life of a human, the mass of ATP generated is approximately equal to the mass of the human body!

As the hydrogen ions pass through the ATP synthase channel, part of the enzyme molecule rotates in a clockwise direction, at approximately 150 times per second, and is capable of generating one ATP molecule for each rotation, as shown in **Figure 2.3.5**. Each turn causes the active site to change and allows the synthesis of ATP. Oxygen eventually combines with hydrogen to form water. Bacteria and chloroplast membranes have a very similar machine to generate ATP.

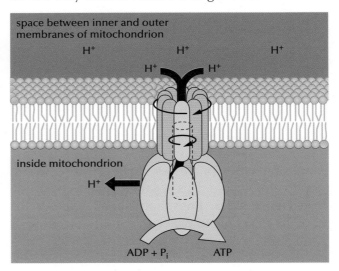

Figure 2.3.5 *Action of ATP synthase*

●GO! Activity 2.3.1 Work individually to:

Restricted response

1. Describe the role of a dehydrogenase enzyme. **1**
2. Give **two** examples of hydrogen carriers in cell respiration. **1**
3. Name the molecule is responsible for the transfer of energy and phosphorylation of molecules within a cell. **1**
4. Resazurin is a dye which changes colour as it gains hydrogen, changing from blue to pink and then then colourless as it picks up more hydrogen. In the presence of a dehydrogenase enzyme, it will readily act as an artificial hydrogen acceptor. $9cm^3$ of yeast suspension was mixed with $1cm^3$ of resazurin. $5cm^2$ of the mixture was added to an equal volume of water (sample A) and the other $5cm^3$ was added to an equal volume of glucose solution (sample B) and the time taken to change colour at room temperature was noted in each case. After 20 minutes, sample A changed from blue to colourless while sample B was pink.

   a) Give a conclusion from these results. **1**
   b) State **three** ways in which the validity of the results could be improved. **3**
   c) Suggest how these results could be made more reliable. **1**
   d) Sample B is a control. Describe how another control could be set up. **1**

Extended response

Give a description of the structure and function of the enzyme ATP synthase. **6**

●GO! Activity 2.3.2 Work in pairs to:

Make a model of the action of ATP as it breaks down to ADP and P_i. With your teacher's permission and giving due care to safety, you might use the novelty explosive strip found in a Christmas cracker to add sound effects!

●GO! Activity 2.3.3 Work in groups to:

Carry out a research project on ATP-dependent luminescence in an animal such as the firefly squid or in photobacteria. Present your findings to the class as a Powerpoint presentation, consisting of five slides.

After working on this chapter, I can:

1. Give a word definition of cell respiration.

2. State that cell respiration occurs in all three domains of living cells.

3. State that the main substrate for cell respiration is glucose.

4. Write a word equation to summarise cell respiration.

5. State that during cell respiration adenosine triphosphate (ATP) is synthesised.

6. State that ATP is synthesised from adenosine diphosphate (ADP) and phosphate (P_i).

7. State the role of a dehydrogenase enzyme.

8. State that the hydrogen released by the action of a dehydrogenase is picked up by a carrier molecule.

9. Give two examples of hydrogen carrier molecules.

10. Give examples of how the energy released during cell respiration is used in living cells.

11. Illustrate the process of phosphorylation by which a phosphate is added to a molecule.

12. State that phosphorylation increases the energy content of the molecule being phosphorylated.

13. Explain why the quantity of ATP stored in a cell is low.

14. State the location of the ATP synthase complex.

15. Explain how the ATP synthase complex works.

16. State that oxygen is the final acceptor of hydrogen ions and electrons to form water.

2.4 Cell respiration 2

Learning intentions

- State that cellular respiration starts in the cell cytoplasm.
- State that in the cytoplasm, glucose is broken down to pyruvate in the process of glycolysis.
- Explain the energy investment involved in the phosphorylation of intermediates in glycolysis.
- Explain the energy pay-off involved in the formation of pyruvate.
- Describe the sequence of events in glycolysis.
- State that pyruvate is broken down to acetyl coenzyme A.
- State that acetyl coenzyme A is transferred to the citric acid cycle.
- State that the citric acid cycle takes place in the central matrix of mitochondria.
- Describe the sequence of events in the citric acid cycle.
- State that ATP is generated, carbon dioxide produced and hydrogen ions released during the citric acid cycle.
- Explain what happens to pyruvate in the absence of oxygen.
- Describe how the electron transport chain operates.
- State that the electron transport chain is located on the inner membrane of a mitochondrion.

- Explain how the flow of hydrogen ions in the electron transport chain is linked to ATP synthesis.
- State that the electron transport chain is the main source of ATP in a cell.
- Describe how oxygen is combined with hydrogen and electrons to form water in the electron transport chain.
- Describe the role of various carbohydrates as respiratory substrates.
- Describe how fats and proteins can also be used as respiratory substrates.

Glycolysis

📖 **Pyruvate**

Important molecule which is formed from glucose after glycolysis.

📖 **Glycolysis**

Initial series of reactions of cell respiration which take place in the cytoplasm, with or without the presence of oxygen.

🔍 **Hint**

Use flow-charts like this to remember the process of glycolysis.

Cell respiration starts in the cytoplasm of a cell when glucose is broken down to form **pyruvate**. This process is called **glycolysis** and does not require any oxygen. The initial reactions of glycolysis use up two ATP molecules as an energy investment phase, to provide energy to convert glucose to intermediate phosphorylated compound. This is followed by reactions which convert phosphorylated intermediate into two pyruvate molecules, which is an energy pay-off phase because four ATP molecules are synthesised. As well as the net gain of two ATP molecules, hydrogen ions are removed from the intermediates by a dehydrogenase enzyme. These hydrogen ions are picked up by NAD to form NADH as shown in **Figure 2.4.1**.

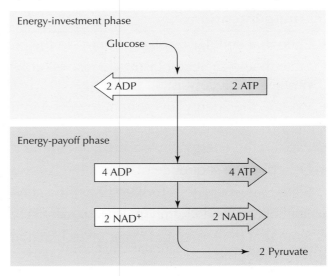

Figure 2.4.1 *The two phases of glycolysis*

📖 **Citric acid cycle**

Cyclical series of reactions operating under aerobic conditions in the mitochondrial matrix.

Citric acid cycle

The second stage in cell respiration, the **citric acid cycle**, takes place in the matrix of the mitochondrion. Pyruvate, produced in glycolysis, is broken down into an acetyl group with the release of carbon dioxide, and more hydrogen ions which are picked up by NAD. The acetyl group combines with **coenzyme A** to form **acetyl coenzyme A** as shown:

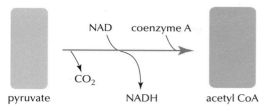

Figure 2.4.2 *Formation of acetyl coenzyme A from coenzyme A (CoA)*

The acetyl group of acetyl coenzyme A joins with **oxaloacetate** to form **citrate**, which then enters the citric acid cycle. As the cycle turns, carbon dioxide and hydrogen ions are released to be picked up by NAD or FAD (to become $FADH_2$). ATP is also produced at one of these stages. The series of reactions which follow eventually regenerate oxaloacetate, as shown in **Figure 2.4.3**.

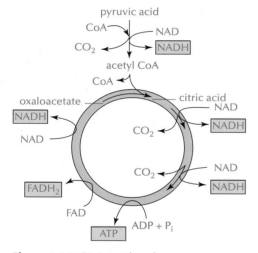

Figure 2.4.3 *Citric acid cycle*

Fermentation

When oxygen is unavailable, **fermentation** takes place. This is much less efficient than aerobic respiration because not all the energy from the substrate, (glucose) is released. The pyruvate formed is converted into **lactate** in animal cells, and **ethanol** and carbon dioxide in plant and yeast cells. Only two ATP are produced from the reactions of glycolysis, as shown in **Figure 2.4.4**.

📖 **Acetyl coenzyme A**

Important intermediate metabolite linking glycolysis to the citric acid cycle.

🔍 **Hint**

Remember the enzymes involved in converting pyruvate to acetyl coenzyme A form a complex.

🔍 **Hint**

A coenzyme is a small, non-protein molecule which combines temporarily with an enzyme, allowing a reaction to proceed.

📖 **Oxaloacetate**

Intermediate compound which joins with acetyl coenzyme A to form citric acid.

📖 **Citrate**

Intermediate compound in the citric acid cycle.

📖 **lactate**

Compound formed as an end-product of anaerobic respiration during strenuous exercise.

📖 **ethanol**

An alcohol produced by yeasts during fermentation of sugars.

📖 **Fermentation**

Breakdown of glucose in the absence of oxygen.

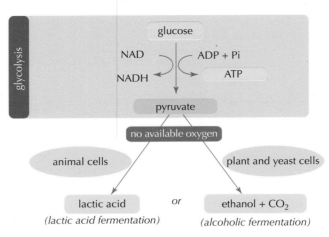

Figure 2.4.4 *Fermentation in the cytoplasm of animal, plant and yeast cells*

Electron transport chain

When oxygen is freely available, the hydrogen ions and electrons picked up by NAD and FAD are transferred to oxygen which is breathed in, and form water molecules:

$NADH/FADH_2$ + oxygen $\longrightarrow$ water + NAD/FAD

Figure 2.4.5 *Formation of water*

📖 Electron transport chain

Series of reactions which occur on the inner membrane of a mitochondrion during aerobic respiration and combine hydrogen and electrons with oxygen to form water.

The $NADH/FADH_2$ do not pass their hydrogen ions and electrons directly to the oxygen, but instead pass these down a chain of enzymes and coenzymes called the **electron transport chain**. These are carrier molecules, which lower the hydrogen ions and electrons step-by-step, from a high energy level present in $NADH/FADH_2$ to a low energy level in water. The carrier molecules are located on the inner membrane of the mitochondria in a cell.

ATP synthesis

Energy, which is released as the hydrogen ions and electrons move down the electron transport chain, is used to pump the hydrogen ions from the matrix to the space between the two membranes of the mitochondria. As the hydrogen ions flow back into the matrix and back across the inner membrane, they pass through the central channel of ATP synthase. This causes the enzyme to rotate, which in turn catalyses the synthesis of ATP from ATP and P_i. This series of events is summarised in **Figure 2.4.6**. Almost all of the ATP made during aerobic respiration is synthesised in this way.

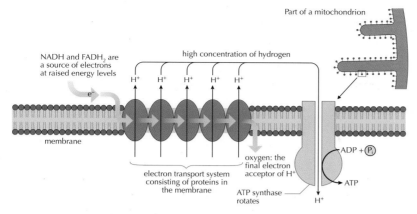

Figure 2.4.6 *The electron transport chain and ATP synthase*

Respiratory substrates

The main substrate for cell respiration is glucose, however other substances may also be used under the correct circumstances. For example, **starch** and **glycogen** may be broken down into the component glucose molecules for cell respiration, whilst the building blocks of fats and proteins can enter the process at various stages as shown in **Figure 2.4.7**.

> ### 📖 Starch
> Carbohydrate stored in plants made up of many glucose molecules.

> ### 📖 Glycogen
> Main carbohydrate stored in animal liver made up of many glucose molecules.

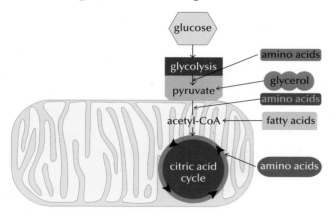

Figure 2.4.7 *Various substrates used in cell respiration and their points of entry*

> ### 🔵 Activity 2.4.1 Work individually to:
>
> **Restricted response**
> 1. State exactly in the cell where the reactions of the citric acid cycle take place.   **1**
> 2. Give the function of NAD in cell respiration.   **2**
> 3. State the **two** products of fermentation in a human muscle cell.   **2**
> 4. Name the final hydrogen and electron acceptor in the electron transport chain.   **1**

5. The following diagram represents a respirometer to measure the rate of oxygen consumption. Soda lime absorbs carbon dioxide.

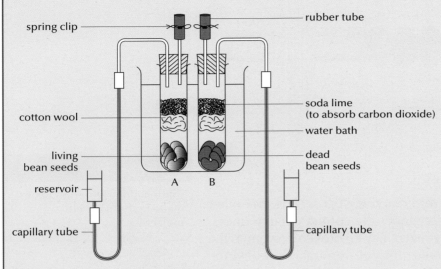

Figure 2.4.8 *Respirometer measuring the rate of oxygen consumption*

a) Give the reason why both tubes A and B are placed in a water bath. **1**

b) State **three** variables which would need to be controlled in this experiment. **3**

c) Tube B is a control experiment. Describe how the contents of this tube differ from those of tube A. **1**

d) Describe how this apparatus would be used to measure the rate of oxygen consumption. **2**

Extended response

Describe the energy-investment and energy-payoff phases in glycolysis. (Labelled diagrams may be used where appropriate.) **6**

GO! Activity 2.4.2 Work in pairs to:

Make a flow-chart of the processes involved in cell respiration with all the labels on separate cards to be placed correctly as a revision tool for yourselves and others in the class.

GO! Activity 2.4.3 Work in groups to:

Research how Sir Hans Krebs discovered the citric acid cycle.

After working on this chapter, I can:

1. State that cell respiration starts in the cytoplasm.

2. State that in the cytoplasm glucose is broken down to pyruvate in the process of glycolysis.

3. State that glycolysis does not require oxygen.

4. Explain why phosphorylating intermediates in glycolysis is termed an energy investment.

5. Explain why the formation of pyruvate is termed an energy pay off.

6. Produce a flow-chart to show the events of glycolysis.

7. State what happens to pyruvate in glycolysis.

8. State what happens to acetyl coenzyme A produced in glycolysis.

9. State the exact location of the citric acid cycle.

10. Produce a flow-chart to show the events of the citric acid cycle.

11. State that the citric acid cycle requires oxygen.

12. List the products released during the citric acid cycle.

13. Explain what happens to pyruvate in the absence of oxygen in animal, plant, and yeast cells.

14. Explain why anaerobic respiration is much less efficient than aerobic respiration.

15. Describe how the electron transport chain works.

16. State the exact location of the electron transport chain.

17. Explain how the flow of hydrogen ions is linked to ATP synthesis in the electron transport chain.

18. State that the electron transport chain is the main source of ATP in a cell.

19. Describe how oxygen combines with hydrogen to form water in the electron transport chain.

20. Name substrates *other* than glucose which can be used in cell respiration.

21. Describe how fats and proteins can also be used as respiratory substrates.

2.5 Metabolic rate

Measurement of metabolic rate

A living organism uses up energy constantly to survive, and the rate at which this energy is used up is called **metabolic rate**. The energy to drive metabolism comes from the breakdown of glucose during cell respiration. Since oxygen is used up and carbon dioxide, water and energy are produced, it is possible to measure the rate of cell respiration by looking at how much:

- oxygen is used up in a given time;
- carbon dioxide is produced in a given time;
- energy (in the form of heat) is released in a given time.

A **calorimeter** is a device which allows the metabolic rate of an organism to be measured. A simplified calorimeter with a human subject is shown in **Figure 2.5.1**.

> ### 📖 Metabolic rate
> A measure of the rate of energy-use in a given time.

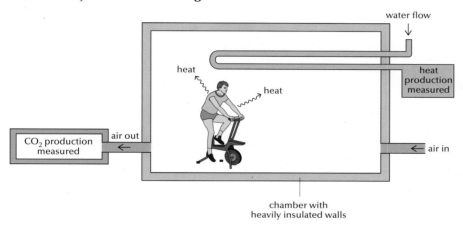

Figure 2.5.1 *Simple calorimeter*

Such measurements can be used to compare the metabolic rates of different organisms at rest as shown in **Figure 2.5.2**.

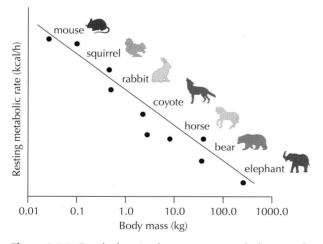

Figure 2.5.2 *Graph showing how resting metabolic rate of some mammals falls as the body mass increases*

Transporting oxygen to cells

The delivery of oxygen to all living cells is vital to supply the reactions of aerobic respiration. Organisms which have a high metabolic rate require a very efficient transport system for this delivery. **Figure 2.5.3** shows the differences in the **anatomy** and **physiology** of the five classes of vertebrates.

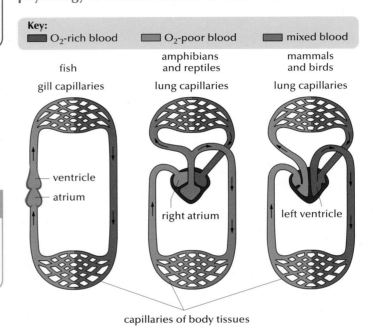

Figure 2.5.3 *Circulatory sytems of different vertebrates*

Fish have a heart with one atrium and one ventricle. The blood flows in one direction only, from the heart to the gills and back to the heart again. It means the body capillaries receive blood at low pressure. This is not a very efficient circulatory system.

Amphibians and reptiles have a heart with one ventricle and two atria. Here, the blood flows through the heart twice for each circuit of the body. The separation of blood that is low in oxygen from blood that is rich in oxygen is not complete. There is some mixing of the blood in the ventricle, but the presence of ridges inside help to direct the two types of blood away from each other. Oxygen-poor blood is sent to the lungs, and oxygen-rich blood is sent to the body tissues. The double circuit of the blood helps to maintain pressure inside the blood vessels.

In mammals and birds, the heart is fully divided into two upper atria and two lower ventricles, and as with amphibians and reptiles, the blood flows through the heart twice for each circuit of the body. In these animals the oxygen-poor blood is kept fully separated from the oxygen-rich blood by a wall running down the middle of the heart. Pressure of the blood is therefore kept high throughout.

Gas exchange

In amphibians, the lungs are small balloon-like sacs which are not very efficient, as the surface area for gas exchange is relatively low. They make use of their moist skin and mouth lining for further gas exchange, as shown in **Figure 2.5.4**.

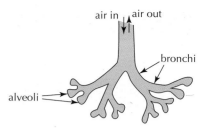

Figure 2.5.5. *Alveoli of reptilian lung*

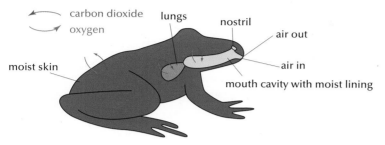

Figure 2.5.4 *Amphibians use moist skin and mouth cavities as well as lungs to exchange gases*

Reptiles have a more developed gas exchange system than amphibians. Their lungs are sub-divided into many small alveolar sacs, which gives a large surface area for gas exchange, as shown in **Figure 2.5.5**.

The lungs in mammals are similar to those of reptiles. There are many millions of alveoli intimately associated with dense beds of blood capillaries, where gas exchange takes place rapidly, as shown in **Figure 2.5.6**. This exchange is aided by the movements of the rib cage and diaphragm.

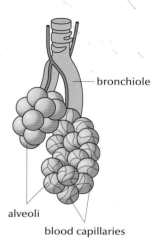

Figure 2.5.6 *The mammalian lung has millions of alveoli for gas exchange*

Birds have the most efficient gas-exchange system of all vertebrates. Oxygen diffuses from the air into the blood through small channels in the lungs instead of blind-ending alveoli. However, the air sucked in does not go directly to the lungs but through the bronchus to large air sacs near the tail end of the bird. At the same time, air already in the lungs moves forward into large air sacs near the head end of the bird. As the animal breathes out, air from the rear air sacs moves into the lungs while air from the top air sacs moves into the bronchi and flows out. In this way, air moves forward through the lungs during both inhalation and exhalation. Gas exchange takes place across the walls of the small channels, as shown in **Figure 2.5.7**.

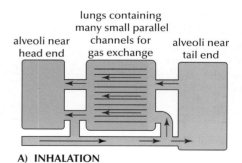

Figure 2.5.7 *Gas exchange in birds. During inhalation, air is drawn in the alveoli in the tail end of the bird (red colour); at the same time, air already in the system is moved forward through the lungs, into the alveoli situated at the head end (grey colour).*

🔍 Hint

Think of the lifestyle of birds and why they particularly need the most efficient gas exchange system of all vertebrates.

🔍 Hint

Notice that during both inhalation and exhalation, oxygen-rich air always moves in one direction through the lungs.

Physiological adaptations of animals for low level oxygen niches

Some air-breathing animals, such as the sperm whale shown in **Figure 2.5.8**, can dive for up to two hours and to depths of 6000 feet. All animals like these have to face the problems of limited oxygen availability, and increased pressure with increased depth whilst searching for food. To cope with these issues, such animals have reinforced airways which allow the lungs to collapse at depth, thereby preventing the absorption of too much nitrogen. Humans experience this excess of nitrogen as 'the bends' and it is potentially fatal. Lung-collapse forces air from the alveoli and this stops the excess nitrogen from being absorbed. The loss of gas exchange with the lung-collapse means the sperm whale has to rely on large oxygen stores in the blood and muscles. Their blood volumes are several times higher than those of land-dwelling animals. The concentration of haemoglobin is about twice that of humans, while the concentration of other oxygen-storing proteins in the muscle is about ten times that of humans. The rate of heartbeat in these animals is also slower at depth, conserving both energy and oxygen demand. Bloodflow to the heart, brains and lungs is temporarily enhanced at the expense of bloodflow to other, less vital organs. Recently it has been discovered that these mammals have additional oxygen-carrying pigments in their brain cells, in addition to the usual haemoglobin in the blood.

Figure 2.5.8 *A sperm whale*

As altitude increases, the atmosphere becomes less dense, which reduces the availability of oxygen for land-dwelling animals. In humans, this can result in a high-altitude sickness with potentially very serious consequences. As humans travel from lower to higher altitudes, the rates of breathing and heartbeat increase. The body can gradually acclimatise to these increases by producing more red blood cells and growing more capillaries. These two physiological changes enable a person to perform at high altitude as well as they did at sea level. Humans who live permanently at high altitude, such as the Tibetan monks shown in **Figure 2.5.9** have increased lung capacity, breathing rates, red blood cell counts, levels of haemoglobin in the blood, numbers of capillaries in the muscles and diameter of blood vessels.

Figure 2.5.9 *Two Tibetan monks*

Before the evolution of photosynthesis, the Earth's atmosphere was oxygen-free. With the appearance of primitive photosynthetic organisms around 3.5 billion years ago, atmospheric oxygen levels rose to around 1%. It took about another 1 billion years for oxygen levels to rise high enough to support the evolution of animals. The current level of oxygen in the atmosphere is 21%, which is enough to allow the development of large land-dwellers such as elephants, which have a high demand for oxygen.

Maximum oxygen uptake as a measure of fitness

The **maximum oxygen uptake** (**VO$_2$ max**) is one way of measuring fitness in humans. Typically, it involves measuring the rate of oxygen consumption during gradually increasing exercise on a motorised treadmill, as shown in **Figure 2.5.10**. The maximum oxygen uptake is an indication of the physical fitness of the person. It is often expressed in millilitres oxygen per kilogram of body mass per minute (ml/kg/min). The VO$_2$ max for marathon running in an athlete is around 80ml/kg/min. As a person gets older, the VO$_2$ max decreases.

> **Make the link**
>
> You will learn more about this in Chapter 1.9 of Unit 1.

> **Hint**
>
> This increase in atmospheric oxygen levels is sometimes referred to as the 'Great Oxygenation Event'.

> **Maximum oxygen uptake**
>
> The maximum volume of oxygen that an individual can utilise during intense or maximal exercise. It is measured as millilitres of oxygen used in one minute per kilogram of body weight.

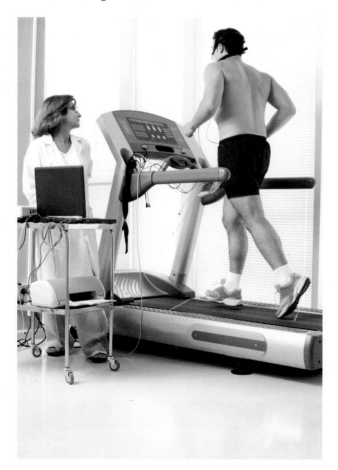

Figure 2.5.10 *Using a motorised treadmill to measure VO$_2$ max*

GO! Activity 2.5.1 Work individually to:

Restricted response

1. State **three** different ways in which metabolic rates can be measured.   3

2. Decide if each of the following statements is true or false. If it is false, give the word(s) which should have been used in place of the word(s) in bold to make the statement correct.   5

   a) A fish has a heart with **two ventricles** with the blood flowing from the **gills** to the **heart**.

   b) Amphibians and **mammals** share a similar heart structure with blood flowing through the heart **once** for each circuit of the body.

   c) Birds have a heart which is divided into **one atrium** and **two** ventricles with blood under **low** pressure.

   d) The skin of a frog must be **dry** to allow gas exchange to occur.

   e) The lungs of a mammal have **few** air sacs which generate a **small** surface area for gas exchange.

3. The table below shows data for the average basal metabolic rate (BMR) of a group of males of different ages measured in megajoules/day (MJ/day).

Age range (years)	Average BMR (MJ/day)
15–24	6·8
25–34	7·1
35–44	6·7
45–54	6·6
55–64	6·5
65–74	5·9

   a) Plot a **bar graph** of these data.   2

   b) State **two** conclusions which are suggested by these data.   2

   c) What is the percentage decrease in the average BMR of 25–34 year olds compared with 15–24 year olds?   1

Extended response

Describe how some animals are adapted for deep diving. (Labelled diagrams may be used where appropriate.)   6

GO! Activity 2.5.2 Work in pairs to:

Constuct tables to show the different adaptations of animals for low-level oxygen niches. Use suitable column headings such as 'Anatomy', 'Physiology' and 'Advantages'.

GO! Activity 2.5.3 Work in groups to:

Prepare a 5-slide Powerpoint presentation to show what VO_2 max is and how it is measured.

After working on this chapter, I can:

1. State three different ways in which metabolic rates can be measured.

2. Understand how these different measurements can allow comparisons of the metabolic rates in different animals to be made.

3. Explain the basic principles of how a calorimeter is used.

4. Explain the relationship between a high metabolic rate and efficiency of the oxygen transport system.

5. Draw fully labelled diagrams showing the anatomy of the heart in fish, amphibians reptiles, mammals and birds.

6. Show the direction of blood flow on these diagrams.

7. Compare the lung arrangements in these animals.

8. Describe the basic operation of the gas-exchange system in birds.

9. Give examples of how animals cope with low levels of oxygen in their environments.

10. Describe how deep diving mammals cope with low oxygen levels and increased pressure.

11. State how over geological periods of time, levels of atmospheric oxygen have changed.

12. Understand how changing levels of atmospheric oxygen relate to maximum body sizes in land-dwelling animals.

13. State what VO_2 max is.

14. Describe how VO_2 max is measured.

15. Describe the relationship between measured VO_2 max and human fitness.

2.6 Metabolism in conformers and regulators

Effect of external abiotic factors

An organism's metabolic rate is sensitive to abiotic factors in its environment. These include:

- temperature;
- salinity;
- pH.

For an organism to survive, these factors need to be within a range which allows the organism's metabolism to function optimally.

Conformers and regulators

Some organisms, such reptiles, amphibians and insects, cannot control their internal environment using metabolism. They must adapt their behaviour, physically moving to and from environments which are more or less favourable. Such organisms are called **conformers**. Conformers usually live in environments which are relatively stable and expend little energy in regulating their internal environment.

Other organisms, such as mammals and birds, have an internal environment which is not dependent on the external environment. They use a variety of metabolic mechanisms to control their internal environments and are called **regulators**. The maintenance of an organism's internal environment independently of the external environment is known as **homeostasis**. Regulators can live across a wide range of environments, with very different temperatures, salinity and pH. However, they expend a lot of energy to maintain homeostasis. It also means they can occupy a wide range of ecological niches.

Figure 2.6.1 shows the effect of changing environmental temperature on the internal (body) temperature of a typical regulator and conformer – here an otter, which is a mammal, and a bass, which is a fish. Notice how the otter's body temperature hardly changes over a wide range of environmental temperatures, whereas the fish's body temperature mirrors the changing environmental temperature.

As they are able to control their internal environment, regulators can survive in a wide range of environments, and exploit diverse ecological niches.

As well as temperature, solute levels in the environment can affect metabolic rate. Most invertebrates which live in salt-water conditions conform to changes in these conditions. For example, a green crab (**Figure 2.6.2**) gains or loses water so that its body fluids match the surrounding water in solute concentration.

> 🔍 **Hint**
>
> What other abiotic factors might impact on an organism's metabolic rate?

> 📖 **Conformer**
>
> An organism which cannot maintain its internal metabolic rate in a changing environment.

> 📖 **Regulator**
>
> An organism which can maintain its internal metabolic rate in a changing environment.

> 📖 **Homeostasis**
>
> General term for the maintenance of body systems, in a stable state which fluctuates above and below optimum levels, using a variety of physiological mechanisms.

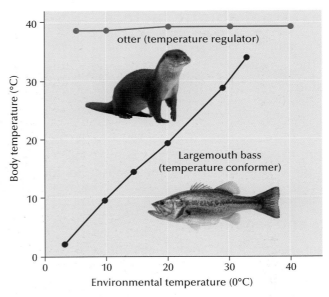

Figure 2.6.1 *Effect of changing environmental temperatures on the body temperature of a regulator and a conformer*

Other organisms can regulate their body solute concentrations and are relatively unaffected by changes in their environments. For example, a freshwater vertebrate such as a goldfish (**Figure 2.6.3**) is actively able to pump out excess water taken to maintain a constant internal solute concentration.

Conformers can only survive changing solute concentrations within a particular range, beyond which the animal will die. However, regulators can survive very wide changes in environmental solute concentration. These differences are shown in **Figure 2.6.4**.

Figure 2.6.3 *The goldfish, an example of an osmoregulator*

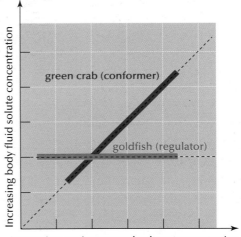

Figure 2.6.3a *Differing responses of a regulator and a conformer to changing surrounding solute concentrations; the dotted line shows the change in body fluids with no regulation as the surrounding solute concentration increases*

Experiments on invertebrates such as shrimp show a relationship between the surrounding pH and metabolic rate. Shrimp are often found in water with a pH of 8. If the pH falls below this value, the metabolic rate decreases, as measured by a slower heartbeat. Most aquatic organisms have an optimum pH of around 6·5 to 8·0, but rapidly growing algae in polluted water, for example, can remove dissolved carbon dioxide to be used in photosynthesis. This increases the pH of the water.

Negative feedback in regulators

Most physiological variables such as temperature are controlled within very narrow limits above or below a **set point**. An important mechanism for this control is called **negative feedback**, and is common in regulators. As the temperature of a mammal's body changes, this is detected by **receptors** which generate nervous signals that go to the **hypothalamus** in the brain. This sends out nervous signals to **effectors**, which respond by activating structures in the skin and muscles. The response attempts to restore the rise above or fall below the set point back to normal. This is summarised in **Figure 2.6.4**.

> ### 🔍 Hint
>
> Consider the indirect effects of climate change on acidity levels in the environment, and how that might affect metabolic rate.

> ### 📖 Set point
>
> Value for some variable which, if departed from, sets in motion a corrective mechanism to reduce the change.

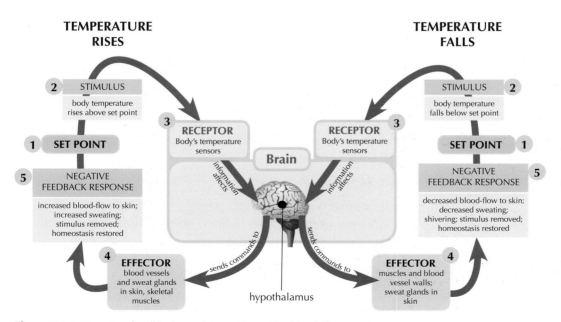

Figure 2.6.4 *Negative feedback mechanism to control body temperature*

📖 Receptor

Specialised group of cells which can detect changes in an animal or plant body or in the environment.

📖 Hypothalamus

Region of the brain associated with monitoring temperature.

📖 Effector

Group of cells which bring about an activity in a muscle or gland, as a result of stimulation from a nervous signal.

Importance of temperature regulation

Since metabolic reactions in living cells are catalysed by enzymes which are sensitive to temperature changes, homeostasis aims to preserve body temperature to as close to its normal set point as possible. This enables the enzymes and cells to work at their optimum efficiency, as shown in **Figure 2.6.5**. Additionally, nerve and muscle cells operate most effectively within a narrow range of temperature.

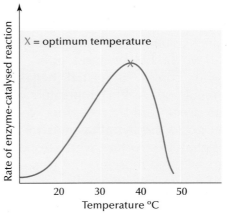

Figure 2.6.5 *How an enzyme-catalysed reaction is affected by temperature*

Many substances pass in and out of a cell by the simple, passive process of diffusion. Increasing temperature increases the rate of diffusion by causing the molecules to move more quickly, as shown in **Figure 2.6.6**.

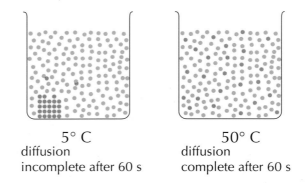

5° C
diffusion
incomplete after 60 s

50° C
diffusion
complete after 60 s

Figure 2.6.6 *Molecules diffuse more quickly at higher temperatures*

Temperature has a profound effect on metabolic rate which is dependent (amongst other factors) on optimum enzyme activity, and efficient diffusion of materials in and out of cells.

Activity 2.6.1 Work individually to:

Restricted response

1. List **three** abiotic factors which can influence metabolic rate.　　　　　2

2. The following words or statements apply either to conformers or regulators. Put the letters under the correct column headings in the table:　　　　　2

   A reptile

   B rely mainly on behaviour to regulate internal environment

   C bird

   D internal environment is independent of external environment

Conformers	Regulators

3. Some data was obtained about the concentration of salt in the body fluids of an aquatic animal (expressed as a percentage) over a range of salt concentrations in the environment (expressed as a percentage). These are shown in the table below.

Salt in body fluids %	Salt in the environment %
1.1	1.1
2.3	2.4
3.2	3.2

   Decide if this animal is a regulator or a conformer and give a reason for your answer, based on evidence from the table.　　　　　3

Extended response

Explain the principal of negative feedback in relation to the control of body temperature.　　6

Activity 2.6.2 Work in groups to:

Each person in a group to **research a conformer or a regulator**. He/she to prepare an A4 poster and and give a short talk to the others in the group.

After working on this chapter, I can:

1. State that an organism's ability to maintain its metabolic rate is affected by abiotic factors.

2. List three different examples of abiotic factors which might affect metabolic rate.

3. State that these factors must fall within a range which allows normal metabolic function.

4. State that some organisms cannot control their internal environment.

5. Give examples of such organisms.

6. Explain why a conformer's internal environment is dependent on the external environment.

7. Explain why a conformer often has a low metabolic rate.

8. Explain why a conformer uses behaviour to maintain its optimum metabolic rate.

9. Explain how a regulator can use its metabolism to maintain its internal environment.

10. Understand the consequences of an organism's ability to regulate its internal environment.

11. State what is meant by homeostasis.

12. Understand why homeostasis in regulators requires energy.

13. State that the range of ecological niches for a regulator is extensive.

14. Explain how regulators and conformers respond to changing abiotic factors.

15. State what is meant by negative feedback.

16. Understand how negative feedback operates in regulating body temperature in mammals.

17. Explain the link between thermoregulation and optimum enzyme function.

18. Explain the link between thermoregulation and diffusion rates.

2.7 Metabolism in adverse conditions

You should already know:

- Adaptation gives an organism an increased chance of survival.
- Adaptations are the result of inherited characteristics which make an organism well suited to its survival in its environment/niche.
- Desert environments are open and arid, with variable temperatures.
- Adaptation can be behavioural, structural, physiological, or a combination of these.

Learning intentions

- State that for any particular organism, an environment may vary beyond the tolerable limits for normal metabolic activity.
- Give examples of how an environment may become intolerable in terms of temperature, water and food availabilities.
- State that one way of surviving adverse conditions is to reduce metabolic rate.
- Explain why reducing metabolic rate increases survival chances in some organisms.
- State that dormancy is part of the life cycle of some organisms.
- State that dormancy may be predictive or consequential.
- Describe that the distinction between predictive and consequential dormancy is not always clear.
- Give two examples of dormancy.
- Explain how each example of dormancy functions.
- State that daily torpor is a period of reduced activity in some organisms which have a high metabolic rate.
- State that one way of avoiding adverse conditions is to migrate.
- Explain the benefit of migration.
- State that migration can occur over very long distances.
- Give examples of specialised techniques used in long-distance studies, including individual marking and tracking.
- State long-distance migration can be carried out by both vertebrates and invertebrates.

- State that migratory behaviour can be influenced by both innate and learned influences.
- Describe studies which can be designed to investigate the effect of innate and learned influences on migratory behaviour.
- State that an extremophile is an organism which can tolerate environments which are potentially lethal to other organisms.
- State that extremophiles are mostly found in the domain archaea.
- Explain the link between the enzymes in extremophiles and their ability to live in hostile environments.
- State that thermophilic bacteria are extremophiles.
- State that some species of thermophilic bacteria generate their ATP by removing high-energy electrons from inorganic molecules.

Surviving adverse conditions

In some environments, conditions of temperature, water and food availabilities may go beyond the tolerable limits in which an organism can survive. Such conditions would cause metabolism to slow down and eventually fail. One strategy organisms use is to slow down their metabolic rate to conserve energy when the cost of maintaining their normal metabolic rate (and thus homeostasis) would be too high. Many organisms have evolved **adaptations** for maintaining homeostasis to survive adverse conditions or to avoid them altogether.

Dormancy

A common strategy used to survive adverse conditions is a reduction in metabolic activity, by entering a period of **dormancy**. There are two types of dormancy:

1. **Predictive**, in which the organism enters dormancy before the adverse conditions begin. This type of dormancy is usually found in environments where changes occur rhythmically, such as with changing seasons. The cue is typically a changing day-length. Some reptiles and amphibians cope with harsh winters by essentially shutting down their body metabolism to an absolute minimum to keep their bodies 'ticking over'. Decreasing daylength is also a strong cue for many plants to shed their leaves and similarly shut down for the winter.

2. **Consequential**, when the organism enters dormancy after the adverse conditions begin. This type of dormancy is usually found when conditions can change unpredictably.

📖 Adaptation

Characteristic which is the result of evolution affording an organism an increased chance of survival.

📖 Dormancy

Period of time in which plants and animals are in a state of low metabolism during conditions which are adverse.

📖 Predictive dormancy

When an organism goes into a state of dormancy before adverse conditions develop.

📖 Consequential dormancy

When an organism goes into a state of dormancy after adverse conditions develop.

The organism is able to react quickly enough to survive in the environment for as long as possible, before entering a fully dormant stage.

It is therefore not always possible to classify an organism as fully predictive or fully consequential in its dormancy.

Hibernation and aestivation

There are two forms of dormancy specific to animals. These are called **hibernation** and **aestivation**.

In harsh winter conditions, hibernation allows animals to survive by reducing their heartbeat, body temperature and oxygen uptake. This is coupled with minimal activity, as shown in **Figure 2.7.1**. These changes reduce the need for energy by lowering the metabolic rate.

In harsh summer conditions where high temperatures and/or lack of water and food can become issues, some animals enter a 'summer dormancy', called aestivation. Land animals may burrow underground to conserve water and avoid the high surface temperatures, as shown in **Figure 2.7.2**.

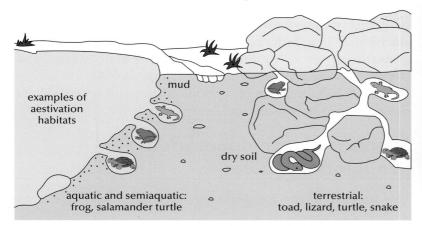

Figure 2.7.2 *Many land-dwelling animals burrow underground to avoid harsh summer conditions*

Daily torpor

When food availability is in short supply, some animals can enter a state of decreased physiological activity, by reducing their body temperature and metabolic rate. Such a state may last a few days or weeks, but **daily torpor** lasts a few hours every day. Animals such as hummingbirds and some mammals, such as mice and bats, which have particularly high metabolic rates, undergo daily torpor as shown in **Figure 2.7.3**.

Figure 2.7.1 *Dormouse hibernating with little or no activity*

Figure 2.7.3 *Hummingbird in a state of torpor on a feeding station outside a family home*

Migration

The bar-tailed godwit, a wading bird shown in **Figure 2.7.4**, makes a non-stop journey of nine days across the Pacific Ocean. Some individuals breed in western Alaska, where there is plenty of food and long days, allowing extended foraging. It spends the winter in Australia to avoid short days and low temperatures.

Figure 2.7.4 *A Bar-tailed godwit feeding*

This type of regular behaviour, which avoids adverse conditions such as very harsh winters, is called **migration**. Migration is not always over long distances. It can be up and down mountains, or vertically through columns of ocean water. A feature of migration is the change of habitat. It allows animals to exploit favourable food and climatic conditions, perhaps only available during limited periods of time.

Specialised techniques used in migration studies

Research into migration would address the following questions:

- When, during their lifecycles, do animals migrate?
- How many times, during their life cyles, do animals migrate?
- Where do migrating animals go?
- What route do they take?
- Why do migrating animals travel to a particular location?
- What triggers migratory behaviour?
- How are factors such as climate, land-use, biodiversity and invasive species changing migratory patterns?

Scientists use a variety of specialised techniques, often in combination, to answer these questions, and with the assistance of modern technology, are more able than ever to collect information.

Ringing

Ringing has been around for over 100 years, and was used initially to follow the movements of birds over long distances. It uses aluminium bands carrying basic information, such as an 'address', unique serial number and contact details, wrapped round the bird's leg as shown in **Figure 2.7.5**.

Figure 2.7.5 *Aluminium ring carrying relevant information attached to birds leg*

Tagging

Alternatives to banding include using dyes and plastic tags. These can be used for mammals and insects as well, as shown in **Figure 2.7.6**.

Radar

Modern radar is very powerful and can plot the height, airspeed, and rate of wing beat of individual animals, such as bats and birds. Sonar, the aquatic equivalent of radar, can similarly track the movement of shoals of fish underwater.

Figure 2.7.6 *Monarch butterfly tagged with a plastic disc*

Radio transmitters

Modern electronic developments make it possible to track animal migration without actually observing an individual. Even tiny insects can be fitted with radio transmitters. This allows their position in space to be followed over a distance dependent on the transmitter and receiver, as shown in **Figure 2.7.7**.

Global tracking

With the increased use of orbital satellite tracking systems, it is now possible to have a transmitter fitted to an animal which will send signals to a global positioning system. This allows scientists to record all kinds of data, without any need for the animal having to be recovered. Some of the transmitters weigh less than 1g, are smaller than a car key and may last for years; some are even solar-powered, as shown in **Figure 2.7.8**.

Figure 2.7.7 *Radio transmitter used to track dragonflies*

Light level monitoring

It is possible to fit an animal with an electronic device which collects light levels over a period of time. The animal has to be recaptured to recover these data and remove the device. The information is used to estimate sunrise/sunset times, which in turn gives an indication of the animal's movements. These devices are particularly useful for animals which spend a lot of time below water, where they cannot easily be tracked by satellites.

Figure 2.7.8 *Eagle fitted with solar-powered radio transmitter linked to orbital satellites*

Innate and learned influences on migratory behaviour

Behaviour which is **innate** is genetically based, so all members of the same species will have this 'hard-wired' into their nervous systems. A newt raised away from water will swim perfectly the first time it is introduced into water. This is clearly a behaviour which does not need practice.

However, behaviour is also influenced by experience, and can be **learned** by observing other animals, trial and error, or being part of a large social group. This type of behaviour is not fixed and the animal must acquire it because it is not an inborn skill.

Migratory behaviour is a product of evolution, so that organisms can adapt to variations in resources. While it is commonly regarded as innate, only some migratory species always migrate, while others may or may not migrate. Owls, for example, may migrate from their environment when their prey is scarce, but not migrate when food is plentiful.

Experimental design

It is sometimes difficult to study migration in real time, so laboratory-based experiments are used to support data obtained by tracking techniques. A great deal of ingenuity on the part of scientists studying the effect of innate and learned influences on migratory behaviour has seen the emergence of very creative experimental designs. Simple observations, such as the intense restless behaviour of some migratory birds kept in captivity during migratory periods (shown in **Figure 2.7.9**) suggest a strong genetic influence on migratory behaviour.

However, the genetically based drive to migrate can be overcome completely. For example, if geese are reared artificially and kept in suitable climatic conditions with plenty of food and mates, they stay where they are reared. Although they are free to leave and follow a 'normal', genetically programmed migratory path, this instinct has been suppressed in favour of ideal conditions for sustained growth and reproduction.

One well-known experiment is based on **displacement**. Here, animals are moved from their familiar environment to a distant location, in order to monitor how they orientate themselves on release. In one very famous early experiment, over 10,000 starlings were relocated by plane from the Netherlands to Switzerland. The displaced young birds were recovered in a south-westerly direction heading towards Spain, but displaced adult birds were recovered heading in a north-westerly direction towards their known wintering sites in north-west Europe. This

Figure 2.7.9 *Some birds get very restless if kept in captivity during a normal migratory period*

suggested that inexperienced birds find their own wintering sites by flying in an inherited direction, but experienced birds can somehow compensate and modify their flight path to arrive at the previously visited wintering sites.

Other experiments involved cross-fostering two related species of birds, such as the herring gull and black-backed gull (shown in **Figures 2.7.10** and **Figure 2.7.11**), which are respectively non-migratory and migratory. A significant number of the non-migratory birds reared by migratory parents showed migratory behaviour, but not as extensively as their migratory parents. This suggests both innate and learned influences are involved. The migratory birds reared by non-migratory parents all migrated, indicating the strength of that genetic influence.

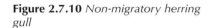

Figure 2.7.10 *Non-migratory herring gull*

Figure 2.7.11 *Migratory black-headed gull*

Displacement experiment

Investigating migratory behaviour by relocating animals and following their subsequent movements after release from the new location.

Extremophiles

Some organisms can thrive in very high pressure, low oxygen, high or low temperature, and high or low pH environments. These organisms are called **extremophiles**. They possess unique enzymes, sometimes referred to as 'extremoenzymes', which enable them to survive in exceptional environments. For example, **thermophilic bacteria** can live in very high temperature water in hot springs.

Extremophiles which live in very hot environments, as high as 100°C, such as hot springs or seabed vents, can generate their ATP by removing high-energy electrons from inorganic molecules. Their metabolism is usually anaerobic, and may involve sulfur-based compounds in their normal metabolism.

The red, flat bark beetle (shown in **Figure 2.7.12**) survives Arctic conditions by producing a kind of anti-freeze cocktail which stops ice-crystals forming in its internal fluids.

The Sahara desert ant is highly active at the hottest period of the day. At this time, surface temperatures can reach 60°C, which is highly inhospitable to its potential predators. They can survive in this temperature by only being active for short periods; they have long legs and move quickly so their body is not in contact with the very hot sand for any length of time.

Thermophilic bacteria

Bacteria which can survive in very high temperatures.

Figure 2.7.12 *Red, flat bark beetle makes its own anti-freeze to survive very cold conditions*

Extremophile

An organism which can survive and grow under extreme environmental conditions.

Hint

Remember that for these organisms, their environments are 'normal'. They appear to be extreme only from our perspective!

The Himalayan jumping spider (shown in **Figure 2.7.13**) lives at heights of nearly 7000 metres, higher than almost every other species. Like ants, these organisms minimise the chance of being predated by operating in such a hostile environment. Winds blow insects up the mountain, which the spider catches for food.

Figure 2.7.13 *The Himalayan jumping spider avoids predators by living in an environment hostile to many of its potential predators*

GO! Activity 2.7.1 Work individually to:

Restricted response

1. A gardener notices some of his plants enter a dormant phase before the onset of winter. They do this every year.

   a) Decide if this is predictive or consequential dormancy and give a reason for your answer. 2

   b) Give two examples of environmental 'cues' which could trigger such dormancy in these plants.   1

2. State which of the following statements apply to aestivation but not to hibernation.

   a) Animals resting in cool, shady and moist places for a relatively short period of time.

   b) Animals spending a long time in dormant conditions, lowering their rate of breathing very considerably.

   c) When surface temperatures climb above a tolerable threshold, some toads can burrow underground until the temperature drops.   1

3. The following data relate to the oxygen consumption of a humming bird in two different states, either torpor or arousal, measured against increasing air temperature.

State 1

Oxygen consumption (cm³/g/hr)	6	3	1	2·5
Air temperature °C	0	10	20	30

State 2

Oxygen consumption (cm³/g/hr)	12·5	10	7·5	2·5
Air temperature °C	0	10	20	30

a) Plot these two sets of data on the same graph.   4

b) Identify which state is that of torpor and which that of arousal.   1

c) State **two** conclusions from these data.   2

Extended response

By means of a named example, explain what is meant by an extremophile. (Labelled diagrams may be used where appropriate.)   4

GO! Activity 2.7.2 Work in pairs to:

Write your own newspaper article outlining how migration can be influenced by both genes and experience.

GO! Activity 2.7.3 Work in groups to:

Produce a 'mind-map' of the usefulness of research into migration. Have 'Research into Migration' as the centre-point and radiate your ideas outwards from this. Use photographs to enhance your mind-map.

After working on this chapter, I can:

1. State that an organism's environment may vary beyond tolerable limits for normal metabolic activity.

2. Give examples of factors which can vary, such as temperature, water, food availability, pH, salt levels, radiation levels.

3. Explain how some organisms can slow down their metabolism to conserve energy.

4. Explain that conserving energy helps preserve homeostasis.

5. State that organisms have evolved adaptations to survive extreme conditions.

6. State that dormancy is one example of an evolved adaptation.

7. Describe the two different types of dormancy: predictive and consequential.

8. Describe hibernation and aestivation in animals.

9. Describe daily torpor.

10. Define what is meant by migration.

11. Explain the benefits of migration.

12. Explain the usefulness of studying migratory behaviour.

13. Know that long-distance migration can be used by vertebrates and invertebrates.

14. Give examples of specialised techniques used in migration studies.

15. Describe innate and learned influences on migratory behaviour.

16. Explain why it can be difficult to design experiments to monitor migratory behaviour.

17. Describe experiments which can be designed to study migratory behaviour.

18. Define what an extremophile is.

19. State that most extremophiles are from the domain archaea.

20. Explain how extremophilic bacteria have special enzymes which allow them to thrive in extreme environments.

21. Describe how one of these extremoenzymes is used in PCR.

22. Explain how some extremophilic bacteria can utilise methane- and sulfur-containing compounds as energy substrates.

23. Give examples of extremophiles other than extremophilic bacteria.

2.8 Environmental control of metabolism

You should already know:

- Microorganisms include bacteria, algae, fungi as well as simple, single-celled organisms.
- Microorganisms can be grown in culture using aseptic techniques, an appropriate medium and the control of other factors.

Learning intentions

- State that microorganisms include archaea, bacteria and some eukaryotes.
- State that microorganisms include species which use a wide variety of substrates for metabolism.
- State that microorganisms produce a wide range of products from their metabolism.
- State that microorganisms are adaptable, easy to grow, and reproduce rapidly.
- State that environmental factors influence microbial growth.
- Describe how conditions for growing microorganisms include controlling temperature, oxygen, and pH levels as well as nutrient supply.
- State that energy for metabolism in microorganisms comes from light or chemicals.
- State that various forms of media can be used to grow microorganisms.
- State that an industrial fermenter allows the growth of microorganisms on a large scale.
- State that growth media can be composed of specific substances.
- Explain why growth media can contain complex ingredients such as beef extract.
- State that many microorganisms can produce all the chemicals required for protein synthesis.
- State that some microorganisms need complex compounds to be supplied for growth.
- Give examples of complex growth compounds needed such as vitamins and fatty acids.
- Explain why conditions for growing microorganisms must be sterile.
- Name four distinct phases of microbial growth.

- Explain what is meant by a generation time.
- Describe the difference between viable count and total cell count.
- State that numbers of cells in a growing culture can be plotted on normal and semi-logarithmic scales.
- Describe how microbial growth can be controlled by adding chemicals to give a required product.
- Give examples of chemicals which can be added to control microbial growth to include metabolic precursors, inducers or inhibitors.
- Describe how secondary metabolism can confer an ecological advantage on a microorganism.

Usefulness of microorganisms

Microorganisms can be prokaryotes or eukaryotes. Prokaryotes include bacteria and archaea while eukaryotes include algae, protozoa and fungi. These organisms are very adaptable, in that they can use a variety of different substrates in their metabolism. The different species of microorganisms often grow on these different substrates, which in turn are major factors in determining their ecological niches. Additionally, the specific metabolism of any one microorganism may allow it to be used in industrial or geochemical processes. For example, some bacteria use methane gas as a respiratory substrate. They have been exploited to detoxify environmental pollution by methane gas from rice paddy fields, refuse dumps, boggy areas, and swamps where this gas is released into the atmosphere.

Microorganisms are widely exploited by humans in both industry and research because they grow very quickly, are adaptable, and are easy to culture. Recent developments have enhanced their use in genetic manipulation so that they can produce a wide variety of useful products. Some of these include antiviral drugs, cancer treatments and insecticides.

Growing microorganisms

The metabolism of microorganisms is easily controlled by altering the conditions under which they are grown. Oxygen and pH levels, temperature, nutrient supply and sterile conditions can all be adjusted, and sophisticated technology has automated this control in many ways, to make the production of a desired product even more efficient.

In order to grow, microorganisms need both an energy supply and a source of nutrients to synthesise the complex chemicals needed. The energy source can be light used in photosynthesis by algae, but almost all microorganisms need a chemical energy source. While many microorganisms can synthesise the chemicals needed

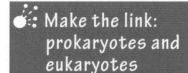

Make the link: prokaryotes and eukaryotes

You have met these terms before in Unit 1, when studying the structure of DNA and how it replicates.

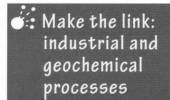

Make the link: industrial and geochemical processes

Think of bacteria and the nitrogen cycle for example.

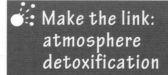

Make the link: atmosphere detoxification

This is one strategy for combating climate change discussed in Unit 3.

for growth from simple compounds such as amino acids, some require more complex compounds added if they are to be grown artificially in the laboratory.

The medium for growing microorganisms can either be solid, in the form of a nutrient-enriched agar jelly or in an equally nutrient-rich broth, as shown in **Figure 2.8.1**.

The nutrient growth medium can contain specific substances, and have complex compounds such as vitamins and fatty acids added or beef extract added, which supplies all of these. An industrial device, called a **fermenter** or **bioreactor**, allows the growth of microorganisms on a large scale as shown in **Figure 2.8.2.** A fermenter allows the medium to be aerated and closely regulated, often by computer control in terms of pH, nutrient levels etc, and other variables.

Figure 2.8.1. *Microorganisms growing on solid nutrient agar*

📖 **Fermenter**

Large container used to grow microorganisms on a commercial scale.

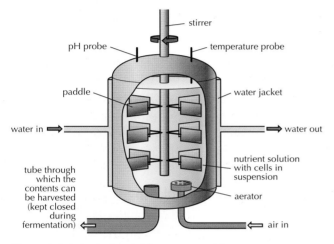

Figure 2.8.2 *An industrial fermenter or bioreactor*

To prevent the contamination of the culture media, special **aseptic techniques** are used to keep conditions **sterile**. One example of an aseptic technique is shown in **Figure 2.8.3**.

Phases of microbial growth

If conditions for growth are suitable, microorganisms can grow very rapidly. Bacteria can double their numbers every 20 minutes. This type of rapid growth, where the numbers/mass of bacteria doubles every cycle of growth is called **exponential**. The time it takes for one cell to divide into two is called the **generation time** or **doubling time**.

There are four distinct phases in microbial growth, as shown in **Figure 2.8.4**. These are:

1. **lag**

2. **log (exponential)**

3. **stationary**

4. **death**

Figure 2.8.3 *Using a Bunsen burner to sterilise a wire loop before and after use*

📖 **Aseptic technique**

A procedure which prevents the environment contaminating a culture or the culture from contaminating the environment.

Sterile

Absence of all forms of living material.

Exponential

When referring to growth, the doubling of numbers or mass every growth cycle.

Generation time

Time taken for one cell to become two.

Doubling time

Time for numbers or mass of an organism to double.

Lag phase

Period when microorganisms are adapting to a new environment and cells synthesise new enzymes.

Log phase

Period when cell numbers double for each growth cycle.

Stationary phase

Period when growth and death rates balance each other out.

Death phase

Period when all the microganisms are dying.

Total count

Measure of all the cells in a culture, both living and dead.

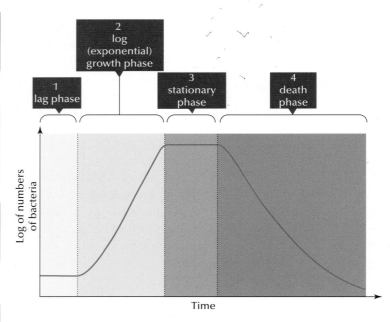

Figure 2.8.4 *Graph of growth pattern in a microorganism over time with numbers shown on a logarithmic scale*

1. In the lag phase, there is very little change in the numbers of the microorganisms. However, their metabolic activity is very high, as the cells adjust to the medium in which they are growing. This might involve new enzyme induction.

2. As long as nutrients are in plentiful supply and the environmental conditions suitable, the microorganisms multiply at a constant rate, so that their numbers increase geometrically during the log or exponential phases.

3. During the stationary phase, nutrients start to get used up, and toxic end-products accumulate as the pH typically begins to change. The number of microorganisms being produced is balanced by the numbers dying so the net growth is zero. During this phase, secondary metabolites are often produced.

4. Eventually, the microorganisms are starved of nutrients or cannot tolerate their own toxic wastes so they die in the death phase.

It is possible to obtain an accurate count of all the cells in a given volume of a microbial culture by removing a drop and placing it on a special counting slide. This gives rise to a **total count**, including both living and dead cells. Only relatively dense samples are useful for this, in excess of 10^7 cells/cm^3.

To obtain the **viable count**, which measures only the living cells, a different strategy is used. One method involves diluting the culture in a known volume of a sterile fluid, and then plating out a fixed volume of the diluted culture. The number of colonies are counted and a calculation done to estimate the number of viable cells. Ideally, between 30 and 300 colonies grow, but if the numbers exceed this, a further dilution of the culture is carried out and the procedure repeated as shown in **Figure 2.8.5**. For example, a count of 30 colonies at a dilution of 1/10,000 gives an estimated viable count of 300,000/cm^3.

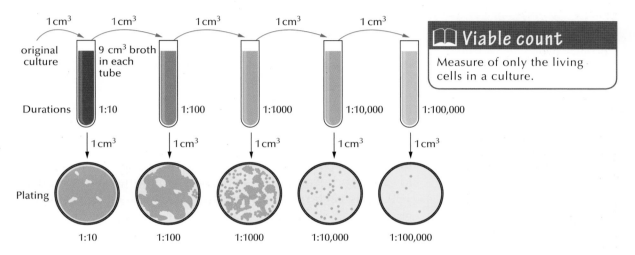

Figure 2.8.5 *Serial dilutions of a microbial culture to give estimated viable counts of microoganisms*

Another strategy is to use a blue dye, which can only penetrate the membranes of dead cells, as shown in **Figure 2.8.6**. Using a counting slide as above, but only recording the unstained cells, gives an estimate of the viable count.

Microbial growth curves

Under suitable conditions, the number of microorganisms in a culture can divide very rapidly, sometimes doubling every twenty minutes. This exponential growth can be plotted in various ways.

One plot simply uses the viable count versus time, as shown in **Figure 2.8.7**. Here the shape of the graph shows the lag and stationary phases. However, the numbers increase so rapidly that it becomes impossible to continue the plot. Additionally, it is hard to see what changes are taking place in the early stages.

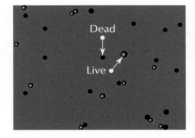

Figure 2.8.6 *Only cells which are dead take up the blue dye; the live ones remain unstained in this viable count method*

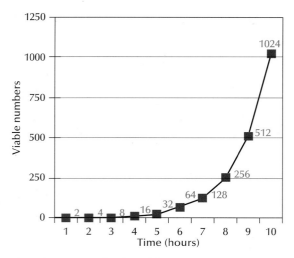

Figure 2.8.7 *Simple plot of the increase in the viable count of a microbial culture versus time*

📖 **Primary metabolism**

Process occurring in exponential phase of microbial growth which produces essential metabolites for cell function.

One way to resolve this is to use a different scale on the y-axis, using special semi-logarithmic graph paper. This is useful when the numbers increase exponentially, because each division on the y-axis is not equal to the next one above; rather the interval for 1 to 10 becomes the same size as the interval from 10 to 100. Thus each regular interval of growth is actually 10 times greater than the one below and gives a straight-line graph. Notice the x-axis has only a restricted range as shown in **Figure 2.8.8** for two regular intervals of growth.

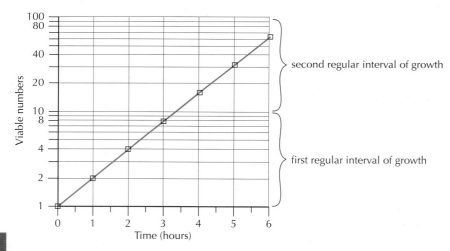

Figure 2.8.8 *Growth of a microorganism plotted on semi-logarithmic graph paper*

📖 **Secondary metabolism**

Process occurring during the stationary phase of microbial growth which produces non-essential metabolites for cell function.

Control of metabolism

Metabolism in microorganisms can be thought of as **primary** or **secondary**. Primary metabolism produces essential metabolites for the normal working of a cell, including nucleic acid bases, amino acids, sugars and fatty acids, as well as important enzymes needed to synthesise these metabolites. These metabolites are formed during the exponential growth phase. Secondary metabolism occurs in the stationary phase of growth, producing metabolites which are species-specific. These can include chemicals, such as antibiotics, of great medical use to humans. While such secondary metabolites are not needed for microbial growth, they can confer advantages on the microorganisms which produce them. For example, the antibiotics they produce may inhibit the growth of competing organisms in the environment, while other secondary metabolites may help the microorganism survive in a different environment.

Since the secondary metabolites of microorganisms can be so useful to humans, their normal pathways of metabolism have been closely studied. Many ways have been found to modify these microorganisms, in order to enhance the production of the secondary metabolites.

Three important compounds can be used to modify normal metabolism to enhance the production of one specific secondary metabolite:

1. **Precursors**

2. **Inhibitors**

3. **Inducers**

Precursors are substrates which form the building blocks of another, more complex compound. For example, the antibiotic actinomycin uses the amino acid tryptophan as a precursor. By artificially adding such precursors, the process of antibiotic production is sped up.

Inducers switch on genes, which in turn produce enzymes, etc. e.g. lactose in the LAC operon.

📖 Precursor

Chemical which forms the basis of a more complex one next in line in a metabolic pathway.

📖 Inducer

Chemical which will promote the activity of a particular enzyme.

Make the link:

You have met inhibitors when studying enzyme function.

🔵 Activity 2.8.1 Work individually to:

Restricted response

1. Give the term used to describe growing microorganisms in the laboratory using aseptic techniques. 1

2. **a)** A nutrient agar plate had $1cm^3$ of a bacterial culture spread evenly over its surface. After incubation, the plate had 35 colonies. If the original culture had been diluted by a factor of 10^5, how many bacteria were present in the original $1cm^3$ sample? 1

   b) A bacterial culture undergoes five complete divisions in two hours. Calculate how long it takes for one division to take place. 1

   c) Calculate how many complete divisions it would take for four bacterial cells to produce 128 cells. 1

3. The graph below shows how the number of microorganisms in a culture changed with time.

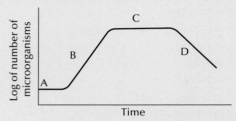

Using the letters A, B, C or D, indicate which description below correctly matches the growth phases on the graph.

a) Number of cells being produced is balanced by the number dying.

b) Number of cells dying is more than the number being produced.

c) Cells are dividing at their maximum rate.　　　　　　　　　　3

4. The following graph shows the cell number of a growing culture of microorganisms against time plotted on semi-logarithmic graph paper.

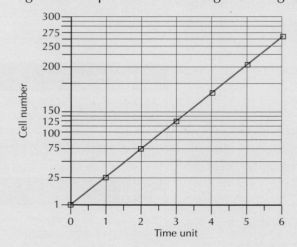

State the number of complete cell divisions shown in this plot.　　　　1

Extended response

Describe three ways of controlling microbial growth to produce a specific product for human use.　　　　6

Activity 2.8.2 Work in pairs to:

Obtain some semi-logarithmic graph paper and download a set of data for the growth of a microorganism over a period of hours. Plot the graph ensuring you use a rule, sharp pencil, label each axis, include the units and join point-to-point. Work out the number of growth cycles.

Activity 2.8.3 Work in groups to:

Brainstorm this chapter. Divide this topic up into suitable sections to match the number of groups. Each group is to construct five multiple-choice questions. Bring all the questions together into one paper, which can then be used for revision.

After working on this chapter, I can:

1. State that microorganisms include archaea, bacteria and some eukaryotes.

2. State that microorganisms include species which use a wide variety of substrates for metabolism.

3. Give an example of an unusual respiratory substrate used by microorganisms.

4. State that microorganisms produce a wide range of products from their metabolism.

5. Give examples of useful products made by microorganisms.

6. State that microorganisms are adaptable, easy to grow and reproduce rapidly.

7. State that environmental factors influence microbial growth.

8. Describe how conditions for growing microorganisms include controlling temperature, oxygen and pH levels as well as nutrient supply.

9. Explain how new technology has made control of microbial growth highly efficient.

10. State that energy for metabolism in microorganisms comes from light or chemicals.

11. Explain why some microorganisms need complex compounds to grow in the laboratory.

12. State that media to grow microorganisms can be solid or liquid.

13. State that growth media can be composed of specific substances.

14. Explain how a fermenter can be used to grow microorganisms on a large scale.

15. State that many microorganisms can produce all the chemicals required for protein synthesis.

16. State that some microorganisms need complex compounds to be supplied for growth.

17. Give examples of complex growth compounds needed, such as vitamins and fatty acids.

18. Explain why conditions for growing microorganisms must be sterile.

19. Describe what is meant by aseptic techniques.

20. Explain what is meant by exponential growth.

21. Describe the four distinct phases of microbial growth.

22. Explain what is meant by a generation time.

23. Describe the difference between viable count and total cell count.

24. Understand that numbers of cells in a growing culture can be plotted on normal and semi-logarithmic scales.

25. Explain the difference between primary and secondary metabolism in microbial growth.

26. Give examples of how humans can use secondary metabolites produced in microbial metabolism.

27. Describe how microbial growth can be controlled by adding chemicals to give a required product.

28. Give examples of chemicals which can be added to control microbial growth to include metabolic precursors, inducers or inhibitors.

29. Describe how secondary metabolism can confer an ecological advantage on a microorganism.

2.9 Genetic control of metabolism

Learning intentions

- State that wild strains of microorganisms can be improved.
- Give examples of ways of improving microorganisms and give examples of how these techniques are used.
- Give examples of mutagenic agents.
- State that mutated forms of a microorganism can back-mutate, making them unstable.
- Describe how some bacteria can transfer genetic information.
- Describe how fungi and yeasts can produce new phenotypes by sexual reproduction.
- Explain how vectors are used in genetic engineering.
- Describe how some genes can remove inhibitory controls or amplify specific metabolic steps.
- Describe a safety strategy which incorporates genes into a genetically modified microorganism, thus preventing that organism from surviving in the external environment.
- Describe how restriction endonucleases are used in genetic engineering.
- Describe how ligases are used in genetic engineering.
- Explain what is meant by marker genes, restriction sites, origin of replication, selective markers and regulatory sequences.
- Explain why yeast cells may be used to avoid polypeptides being folded incorrectly, or lacking post-translational modifications.

Techniques for improving wild strains of microorganisms

Scientists have found many wild strains of microorganisms which have desired phenotypes, typically by expressing a product which is useful to humans. For example, yeasts have been used for hundreds of years to produce wine, by converting sugar to ethanol. However, by using modern techniques, the genes which control this function have been fully mapped, opening up the possibility of modifying the yeasts to make the production of biofuels very efficient and cost-effective. Ethanol is essentially toxic when it accumulates in a yeast culture but by genetic manipulation, strains can be produced which will tolerate high levels of ethanol, making them very useful for biofuel production.

Three important techniques used in genetically improving microorganisms are:

1. **Mutagenesis**
2. **Selective breeding**
3. **Recombinant DNA technology**

📖 Mutagenesis

A process where the genetic information of an organism is altered in a stable manner, resulting in a mutation.

📖 Selective breeding

Artificially choosing organisms which have a desirable trait and using them as breeding sources.

📖 Recombinant DNA technology

Manipulation of DNA by humans.

Mutagenesis

Agents such as ultra-violet radiation and mustard gas are known mutagens, which cause changes in the DNA of organisms. Such changes are known as mutations. The cells affected may then make an altered form of a metabolite, which might be useful to humans and can thus be collected. Sometimes a mutant form of a microorganism is actually an improvement on the normal wild strain, and this new strain can be isolated and cultured up for use industrially. Microorganisms used in commercial contexts often have had deliberate changes made to their DNA at specific sites. Mutant strains of microorganisms can often back-mutate to their original genetic makeup, making them unstable in the long term and no longer capable of producing the desired metabolite. To deal with this, cultures of the microorganism are constantly made anew, to ensure the mutated form is always available.

📖 Conjugation

Process, similar to sexual reproduction, by which genetic material is transferred horizontally between different strains of bacteria.

Selective breeding

Although bacteria do not usually reproduce sexually, they can still produce variants by transferring genetic material between different strains. One important way of doing this is called **conjugation** shown in **Figure 2.9.1**.

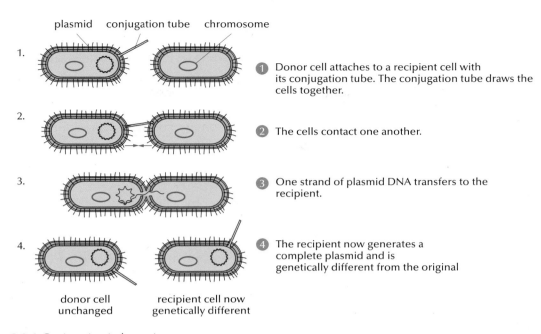

1. Donor cell attaches to a recipient cell with its conjugation tube. The conjugation tube draws the cells together.

2. The cells contact one another.

3. One strand of plasmid DNA transfers to the recipient.

4. The recipient now generates a complete plasmid and is genetically different from the original

Figure 2.9.1 *Conjugation in bacteria*

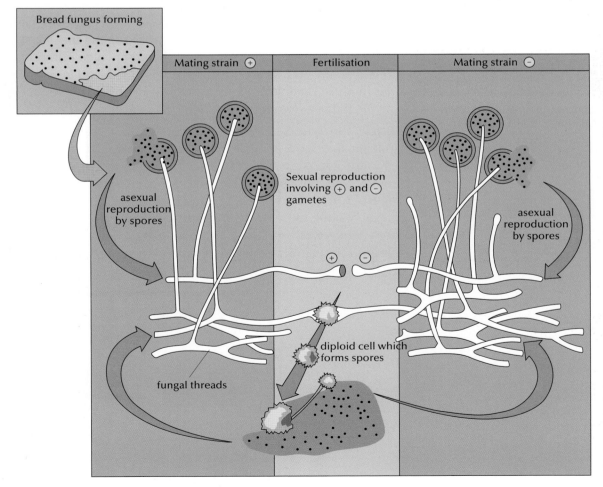

Figure 2.9.2 *Bread fungus can reproduce asexually and sexually*

Yeasts and fungi are capable of reproducing both sexually and asexually. An example is shown in **Figure 2.9.2**. Asexual reproduction essentially produces clones of cells, all genetically identical; however, sexual reproduction gives rise to variation in the offspring. By continually selecting out those offspring produced as a result of sexual reproduction with desirable traits, it is possible to produce new strains from two or more wild strains. These will have a combination of desirable traits from the wild strains.

Recombinant DNA technology

Recombinant DNA technology allows genetic material to be transferred between different organisms which may not be of the same species. Such a transfer can be from a human cell's DNA to a bacterial cell's DNA, for example. The transfer is brought about by a **vector** such as a plasmid, virus, or a piece of artificial chromosome. This forces the bacterium to act as a host. In so doing, it may produce a metabolite, such as insulin. It normally would not do so, but because of the transferral of DNA, it produces a substance which is useful to humans. The gene(s) inserted affect a stage in a metabolic process, so that a metabolite is produced in large quantity. This could be achieved by eliminating the effect of an inhibitory control, or amplifying the specific step. The metabolite is usually very easy to harvest, which is important commercially. As a safety precaution, it is possible to insert genes into the host organism, which renders it unable to reproduce or survive in the external environment.

Special enzymes called **restriction endonucleases** recognise specific areas on a DNA strand called **restriction sites**. The endonucleases cut the DNA into fragments which have 'sticky ends', with unpaired nucleotides at each end, as shown in **Figure 2.9.3.** The same enzymes can be used to cut open a plasmid at the same restriction site, so that the sticky ends of the host gene exactly fit into the two exposed ends of the cut plasmid. To combine the host gene and the cut plasmid another group of enzymes is used, called **ligases**.

📖 Vector

Plasmid, virus or piece of artificial chromosome which carries a gene or genes for insertion into another organism.

📖 Restriction endonuclease

Enzyme which can split DNA at a specific position.

📖 Restriction site

Short sequence found on both strands of a DNA molecule which can be recognised by an endonuclease.

📖 Ligase

Enzyme which is able to join two detached strands of DNA.

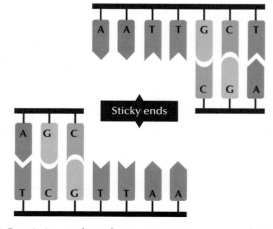

Figure 2.9.3 *Restriction endonucleases expose two unpaired sticky ends*

The modified plasmid cell can be inserted into a bacterial cell. The bacterial cell multiplies rapidly and produces many copies of the human gene and, in turn, synthesises a large quantity of this hormone. Usually the recombinant plasmids carry some **marker genes**, which allow the modified host cells that have taken up those plasmids to be identified. Two common marker genes alter the host cell, making it resistant to antibiotics or making it glow under special lights if it has incorporated the recombinant plasmid. Further, the plasmid will have incorporated an **origin of replication** which consists of genes controlling self-replication of the plasmid DNA, as well as regulatory sequences. These genes usually decrease inhibition or modify metabolic pathways to increase particular metabolites, thus ensuring that multiple copies of the wanted plasmid are made within the bacterial cell, thereby increasing the yield. **Figure 2.9.4** shows how a recombinant plasmid is made:

📖 Marker gene

Section of nucleic acid that allows a host cell which has taken up a modified plasmid to be identified.

📖 Origin of replication

Particular part of the DNA molecule where replication is started.

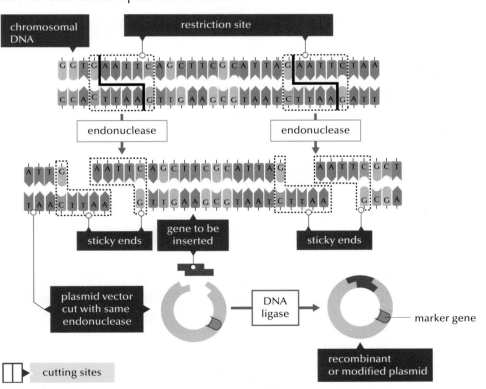

Figure 2.9.4 *Recombinant gene technology*

Recently, mice have been bred which have an extra chromosome which has been made in the lab. Such artificially produced chromosomes have been likened to a shuttle which can act as a vector carrying many more genes than a virus or plasmid.

Another useful organism for use as a vector is yeast, which can be genetically modified to produce proteins normally found in eukaryotic cells. One reason for using yeast instead of bacteria is that the proteins formed by recombinant bacteria do not undergo post-translational modification and do not fold properly to form the correct 3-dimensional structure and therefore do not function properly. The process of genetically modifying a yeast plasmid is shown in **Figure 2.9.5**.

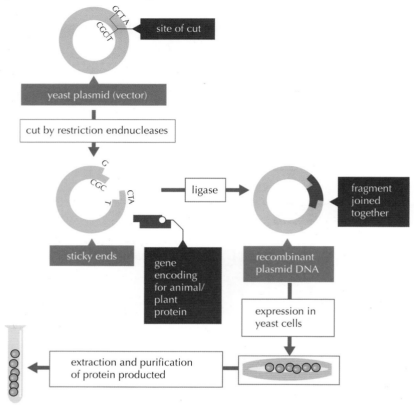

Figure 2.9.5. *Using a yeast plasmid as a vector*

Activity 2.9.1 Work individually to:

Restricted response

1. While mutations occur naturally, they can also be induced artificially. State two ways in which this can be brought about, and give one example of each. 2

2. Mutations in microorganisms can result in the lack of an inhibitory control mechanism. Explain how this can be exploited for human benefit. 1

3. Select one of the following which is **not** a feature of a pure strain of a microorganism. 1

   A Genetically stable.

   B Grows easily in lab conditions.

   C Produces an easily harvested metabolite.

   D Will not overproduce a wanted metabolite.

4. State **one** way in which the risks involved in working with genetically modified microorganisms can be reduced. 1

5. Give an example of a marker gene which can be added into a modified plasmid. 1

Extended response

Describe how a bacterium such as *Escherichia coli* can be modified to produce human insulin. 6

GO! Activity 2.9.2 Work in pairs to:

Create a flow chart to show how a yeast such as *Saccharomyces cerevisiae* can be used as a vector in genetic engineering.

GO! Activity 2.9.3 Work in groups to:

Produce a set of five Powerpoint slides with all the important terms in this chapter. Then divide the class into teams, each one of which nominates a person to sit at the front of the lab, back facing the screen. For each team in turn, a term is put out, so the team can see it but not person at the front. He/she has to work out the term only from the prompts given by his/her team within an agreed time limit. The team with the highest score at the end wins.

After working on this chapter, I can:

1. State that wild strains of microorganisms can be improved by genetic manipulation.

2. Give examples of useful products produced by genetic engineering.

3. Give three examples of genetic manipulation techniques.

4. Describe how these techniques are used.

5. Give examples of mutagenic agents which are physical or chemical.

6. State that mutated forms of a microorganism can back-mutate, making them unstable.

7. Describe one way in which this instability can be resolved.

8. Explain how bacteria can produce genetic variants naturally by conjugation.

9. Describe how fungi and yeasts can produce new phenotypes by sexual reproduction.

10. State that asexual reproduction produces clones.

11. Explain how vectors are used in genetic engineering.

12. Give examples of vectors used in genetic engineering.

13. Describe how some genes can remove inhibitory controls or amplify specific metabolic steps.

14. Describe a safety strategy which incorporates genes into a genetically modified microorganism, which prevents that organism from surviving in the external environment.

15. Describe how restriction endonucleases are used in genetic engineering.

16. Describe how ligases are used in genetic engineering.

17. Understand what is meant by marker genes, restriction sites, origin of replication, selective markers and regulatory sequences.

18. Explain why yeast cells may be used to prevent polypeptides from being folded incorrectly or lacking post-translational modifications.

2.10 Ethical considerations in the use of microorganisms

Learning intentions

- Be aware that increasing use of biotechnology raises ethical issues.
- Describe the hazards involved in using genetically modified microorganisms.
- Describe the ways to control the risks involved in using genetically modified microorganisms.

Ethical issues

Biotechnology involves microorganisms, and so the associated ethical issues related to it concern any potential microorganism's effect on the environment, whether directly or indirectly. Since microorganisms are found everywhere in the Earth's biosphere, the effects of mismanaging this aspect of biotechnology can be extensive and potentially irreversible.

Hazards

Microorganisms exist in colossal numbers in the environment. Microorganisms can multiply rapidly, and quickly be dispersed into the environment by natural means such as wind and water. In this manner, they are able to reach a huge range of ecological niches. This poses issues for organisms already occupying those niches. Microorganisms, being haploid, also express all mutations, and are capable of transferring genetic material horizontally to other similar microorganisms. A genetically modified microorganism, even if dead, can transfer genes which have been engineered in the laboratory. The hosts which acquire these genes may then adapt and survive in a new environment. This can seriously disturb the ecological balance in that new environment. Additionally, the random uptake of new genes can lead to new strains of a pathogen emerging. These may have the ability to resist antibiotics, or evolve to become particularly virulent and able to infect new hosts. Gene transfer may see unwanted metabolites being produced which could damage plants or build up toxins in the environment.

While microorganisms can be genetically modified to meet a specific need, an exact understanding of all metabolic pathways is still not known. This means if another species takes up the modified genes, the outcome cannot be predicted with certainty. The use of 'transgenic' species, which have modified genes for pathogenic properties, is a common approach now to researching ways to combat pathogens. However, this has seen the emergence of new strains of pathogens, with increased power to cause disease. The danger is that these might escape the strict controls needed to prevent them from having unpredictable impacts on humans, and other species.

Minimising risks

Genetic modifications of microorganisms do occur naturally, and mutant forms emerge from time to time as well. The environment will naturally select out variants which are best adapted to survive. Spontaneous mutant forms of yeasts have been used widely in beer manufacture for many years. However, recombinant microorganisms generally are not well adapted to survive in a new environment. In fact, microorganisms cultured continuously in lab conditions often lose their ability to survive in their natural environment. There is now a 'Scientific Advisory Committee on Genetic Modification [SACGM]', which gives clear directives to scientists (and others) on all aspects of working with genetically modified microorganisms. They have developed a rigorous risk-assessment procedure, which is contained in the 'Genetically Modified Organisms 2014 Regulations'. Their goal is to attempt to quantify the risk, and evaluate their probability of happening. The SACGM have devised categories of risk from 1 through to 4.

Additional measures to minimise risk include:

- Physically engineering the environment, using specialised labs, culture equipment, negative pressures, and atmospheric filtration.
- Chemically disinfecting areas.
- Use of extreme heat to sterilise materials and destroy microorganisms after their use has ceased; this can also be used to render waste materials safe for disposal.
- Use of microorganisms which have been genetically disabled, outwith laboratory environments. This is achieved by ensuring that they can only obtain useful nutrition from a lab-based envirnoment.
- Radiation to maintain sterile conditions.

> **Hint**
>
> 1g of human faeces can contain 10^9 *Escherichia coli*!

> **Hint**
>
> Human activity can also allow microorganisms to come into contact with a new environment, as happened when a university flushed out all its aquarium tanks into water which ended up in the nearby sea some years back.

> **Hint**
>
> In 2003, scientists found a way of inhibiting the activity of a gene in the bacterium which causes tuberculosis in humans, and created a strain which was many times more pathogenic that the native strain.

> **Hint**
>
> Experiments with bacteria and yeasts have shown that after several generations of culturing on an artificially enriched media, they are unable to survive in their natural habitat, where nutrients may be in shorter supply.

Activity 2.10.1 Work individually to:

Restricted response

1. Complete the following paragraph using the word bank provided by inserting the correct word into the spaces indicated. 5

> ethical biotechnology haploid genetic rapidly antibiotics environment mutations hazards virulence humans

> ____ allows scientists to manipulate the ____ material found in microorganisms. However, such procedures can raise ____ issues associated with potential ____. Since microorganisms can multiply easily and ____ and are highly adaptable, there is the danger of them escaping into a new ____ and colonising it. Because microorganisms are ____, any ____ will be expressed and if these confer, for example, increased ____ or resistance to ____, there is a danger to other species, including ____.

2. State **three** ways in which scientists can reduce the potential risks associated with biotechnological work. 3

Activity 2.10.2 Work in pairs to:

Organise a debate on the issue, 'should we modify microorganisms?' Decide who will argue for modification and who will argue against. Each should prepare an argument. Hold the debate with the rest of the class as the audience. Your teacher may act as chair and take votes for or against the issue.

After working on this chapter, I can:

1. Describe how using biotechnology presents certain ethical issues.

2. Describe a range of potential hazards associated in using genetically modified microorganisms.

3. Describe a range of ways to minimise the risks associated with using genetically modified microorganisms.

3.1 Sustainable production and supply of food

• Food supply and food security • Global demand for food • Sustainable food production • Increasing agricultural food production • Fertilisers • Competition • Livestock

3.2 Photosynthesis

• Process of photosynthesis • Biochemistry of photosynthesis • Photolysis and the Calvin cycle

3.3 Plant productivity

• Measurement of plant productivity • Leaf area index • Harvest index

3.4 Plant and animal breeding

• Improvement of productivity • Field trials • Breeding animals and plants • Outbreeding • Inbreeding • Monohybrid cross • Animal and plant improvement using gene technology • Genome sequencing • Transgenic animals • Transgenic plants

3.5 Crop protection and livestock welfare

• Crop protection • Weeds • Chemical and biological control of pests • Control of plant diseases • Fungicides • Disadvantages to chemical control methods • Integrated pest management • Animal welfare

3.6 Symbiotic and social interactions between organisms

• Symbiosis • Parasitism • Mutualism • Endosymbiont theory • Social behaviour • cooperative hunting • Social hierarchy • Co-operative defence against predation • Social insects • Keystone species • Behaviour of primates • Parental care • Reinforcing social structure • Ecological niche • Resource distribution • Taxonomic group

3.7 Biodiversity

• Mass extinction • Importance of biodiversity • Threats to biodiversity • Habitat islands • Habitat fragmentation • Habitat corridors • Human exploitation • Bottleneck effect • Introduced, naturalised and invasive species • Climate change

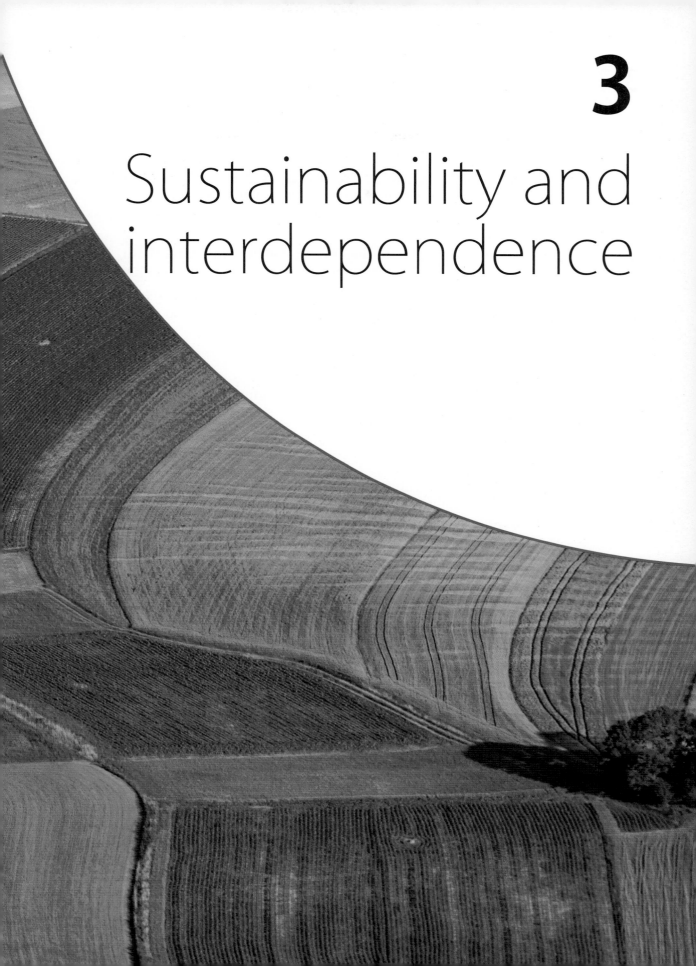

3

Sustainability and interdependence

3.1 Sustainable production and supply of food

You should already know:

- Food chains begin with green plants known as producers, which are able to convert sunlight energy into chemical energy through the process of photosynthesis.
- In transfers of energy from one level to the next in a food chain, 90% of the energy is lost as heat, movement or in undigested materials.
- Fertilisers contain nutrients required for plant growth, such as nitrogen, potassium, and phosphate; the use of fertilisers can increase plant growth.
- Monoculture is an intensive farming practice where one genetic variety of crop is grown over a very large area.
- Chemical and biological methods of protecting crops from pests and diseases.

Learning intentions

- Food chains begin with green plants known as producers, which are able to convert sunlight energy into chemical energy through the process of photosynthesis.
- Explain why there is an increased global demand for food production.
- State the importance of sustainable food production.
- Describe some factors which can affect plant growth and productivity.
- Explain how productivity of a crop may be increased in a limited area.
- Describe the reasons why livestock produce less food per unit area than crops.

Food supply and food security

Food supply

Food supply

Availability of food.

The term **food supply** refers to the availability of food to individuals. In developed countries, the majority of food is accessed through shops, with the price set by retailers. Food supply must be able to meet consumer demand. If the supply is low and demand is high, prices rise and fewer people have access to the available food.

In developing countries, food is often accessed through local markets or food aid agencies. Demand for food here is much greater than the available supply, meaning many people will go hungry if they cannot afford local prices, or travel to reach a food aid centre.

Figure 3.1.1 *Food is accessed through supermarkets and shops in developed countries*

🔍 **Hint**

Food banks, found in many urban areas, improve food security.

Figure 3.1.2 *In developing countries, food is accessed through local markets or aid distribution centres*

Food security

Food security means that at all times, people are able to physically find a source of high-quality, nutritious food, and can afford to buy it. This allows people to sustain a healthy lifestyle.

📖 **Food security**

Location and affordability of food.

Figure 3.1.3 *The nutritional value of food is increased if there is access to clean drinking water*

Achieving food security depends upon:

1. **Food access** – people being able to access a range of high-quality foods that comprise a balanced diet

2. **Food availability** – enough high-quality food being available to people from local producers, food agencies or imports

3. **Food use** – available food being maximised by healthful preparation techniques, in wide variety, combined with the ready availability of clean drinking water

4. **Food sustainability** – people having access to quality food over long-term timescales

Global demand for food

Demand for increased food production

The global demand for food is increasing as the global population continues to rise. The demand for food now currently outstrips the supply. If the current pattern of birth rates exceeding death rates continues, the demand for food will have increased significantly by 2050. At this point, the human population is predicted to have reached 9 billion people.

People living in Europe and America have strong economies which produce thriving food manufacturing systems. The end result of this is that the existing food supply is put under further demand as it struggles to satisfy the appetites of the market. As economies continue to emerge and develop globally, areas which had traditionally consumed mostly staples, such as rice, maize and barley are now able to buy higher-value items such as meat. Providing feed for livestock puts a demand on food supply, and meat is costly to produce in terms of energy.

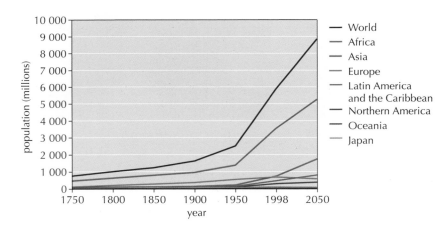

Figure 3.1.4 *Changes to the world's human population from 1750*

Concern for food security

People in developed countries are secure that food will be available to them for purchase at all times. There is also an expectation of there being a wide range of fresh, high-quality food to choose from. As a result of this concern for food security, there is an increased demand on food producers to be able to continually supply supermarkets all year round. As demand on food production systems continues to rise, it is essential that these are developed sustainably for the future.

Sustainable food production

The aim of producing food in a **sustainable** way is to increase crop yields, whilst conserving the natural resources upon which agriculture depends, wherever possible. This approach works only if it directly involves farmers, farm workers and the local community. Sustainable food production systems make a valuable contribution to both local economies and sustainable lifestyles. By using fewer chemicals such as fertilisers and pesticides for growing crops, the biodiversity of animal and plant life is protected. The healthy growth of crops depends upon soil being in a good condition.

Sustainable food production practices maintain the mineral content of soil and improve drainage through maintenance of the soil structure. As sustainable food production systems are generally more labour intensive, this provides opportunities for local employment. Another benefit is the production of good quality food, which people may prefer to purchase, secure in the knowledge that it has been produced in a sustainable way without harming the environment.

Factors contributing to sustainable food production

- Efficient use of available water for crop irrigation.
- Soil management practices that reduce water logging and **soil erosion**.
- Optimum number of grazing animals per area of land (stocking density) to prevent over- or under-grazing of pasture.
- Optimum levels of fertiliser applied to crops to maximise growth.

📖 Sustainable

Conserving an ecological balance by avoiding the depletion of natural resources.

📖 Soil erosion

Loss of soil particles from a field due to wind or rain, resulting in poor soil structure.

Figure 3.1.5 *Ploughing soil to the correct depth allows oxygen into the soil; this promotes strong root growth, and reduces waterlogging*

- Moving away from chemical fertilisers which have a high energy cost to produce, to lower-energy alternatives such as seaweed and cattle manure.

- Prevention of the removal of nitrogen, potassium and phosphorus from the soil through practices such as monoculture.

- Reduction of use of fossil fuels within production, used for powering tractors and other machinery.

- Use of high-yielding genetic varieties (cultivars) of crops, in conjuction with genetically modified (GM) crops. These are pest- and disease-resistant, which reduces the need for chemicals.

Increasing agricultural food production per unit area of land within commercial systems
High-yielding cultivars

Knowledge of factors which affect plant growth and photosynthesis will increase the yield per hectare of either a growing crop, or livestock. (Limiting factors in photosynthesis are discussed in Unit 1.)

> **Hint**
>
> Remember, an increase in the rate of photosynthesis results in an increase in dry mass (yield).

> **Cultivar**
>
> A genetic variety of plant that has been selected and maintained in cultivation through agricultural practices.

The saleable yield of grains from a cereal crop can be increased from a small area of land by selecting a genetic variety, or **cultivar**, which has fewer large grains per head, and a short straw (also called a stem). The short straw avoids energy being directed into the growth of a part of the plant with little value, and more energy being directed into the grain. Cereals with a short straw are less likely to bend and break in bad weather, dropping valuable seeds onto the ground.

Fertilisers

The application of nitrogen fertiliser can increase the growth of plants, resulting in dry mass being greatly increased. Nitrogen fertiliser is often applied to grass, in order to increase the food available to sheep and cattle within that area of land. Nitrogen fertiliser is expensive, and can cause algal blooms in waterways when washed out of the soil by rain water.

Figure 3.1.6 *Genetic cultivar of wheat which has larger grains per head*

Cereals are able to produce heavier grains per unit area if a compound fertiliser is applied containing the elements nitrogen, phosphorus and potassium.

Competition

The productivity of a growing crop can be increased if competition is reduced. Growing crop plants such as wheat compete with each other for space, water and nutrients. This is called **intraspecific competition**, and can be reduced, if seed spacing during the planting of the crop is done correctly.

Growing plants can also compete with different species of plants within the field (such as weeds) for available resources. This is called **interspecific competition**, and can be reduced through physical removal of the weeds, or by applying chemical herbicides.

Pests such as greenfly pierce leaves and stems with sharp mouthparts, feeding on the glucose within phloem tubes. The plant must compete with the pest for available glucose. When a fungal disease affects the leaves of a plant, fungal threads (hyphae) grow into the leaf cells competing with the plant for cell resources such as glucose and water for growth.

Livestock

Meat is produced from domesticated animals such as cattle, sheep, pigs and poultry. Herbivores occupy the second trophic level in food chains. Only 10% of energy from grass is passed onto the cattle and assimilated into saleable muscle tissue. Most energy is lost as heat. As a result, livestock produce less food per unit area than crops.

📖 Intraspecific competition

Competition between members of the same species for the same resources.

📖 Interspecific competition

Competition between members of different species for resources such as water and minerals.

Figure 3.1.7 *Cattle produce less food per unit area than food crops such as cereals*

⚗️ Make the link

Farmers must calculate a 'gross margin' for each food enterprise. If growing a crop of wheat to sell to the baking industry, the farmer must calculate the cost of growing and harvesting the crop. He would begin by listing the fertiliser and seed costs, the labour costs for tractor drivers per hour, the hire of a combine harvester, together with the cost of pesticides and fungicide sprays. The total cost of growing and harvesting the crop is subtracted from the money gained through the sale of the grain. This gives the gross margin. A positive gross margin means the farmer has made a profit, and a negative gross margin means he has made a loss.

⚗️ Make the link

The productivity of a crop is the mass of carbohydrate produced during photosynthesis, which is converted into cell components through assimilation, described in Unit 3.3.

🔵 Activity 3.1.1 Work individually to :

Restricted response

1. a) Explain what is meant by the term 'food security'.   2
   b) Identify two factors which may lead to an increase in the global demand for food.   2
   c) Give reasons why is it important to develop methods of sustainable food production. 1
   d) Explain how the following can increase food production.
      i. the reduction of competition;
      ii. the use of fertilisers.   2

(continued)

2. a) Information about UK cereal production is shown in the table below.

Year	Cereal crop	Yield (millions of tonnes)
2011	Barley	1·4
2012	Barley	1·7
2011	Wheat	2·1
2012	Wheat	1·3

   i. Calculate the percentage *increase* in barley yield, between 2011 and 2012.   1

   ii. Calculate the percentage *decrease* in wheat yield, between 2011 and 2012.   1

   iii. Calculate the total average yield of all cereals in the UK, over the year period.   1

b) Information about the total area of crops grown in UK is shown below.

Crop	Year	Area grown ('000, hectares)
Potatoes	2000	304
	2001	342
	2002	499
	2003	384
	2004	590
	2005	356
Maize	2000	112
	2001	108
	2002	207
	2003	143
	2004	120
	2005	198

Present this information as a line graph.   2

Extended response

Describe the ways in which food production can be made more sustainable.   6

(GO!) Activity 3.1.2 Work in pairs to:

1. Using information contained in this chapter, make an informative leaflet explaining what 'food security' means, and why it is important.

2. **Read** the following paragraph and **present** the information given below as a table:

   Soil management is an important factor in maximising the productivity of crops. Research has shown that a pH of 6·0 gives a maximum spring barley yield of 100%. Soil pH levels of 5·3 and 5·4 give yields of 60% and 75%, respectively. A reduction in maximum yield occurs at pH 6·3, to 90%; maximum yield is 100% at pH 6·0. A soil pH of 6·0–6·2 helps to maintain phosphate and potassium levels for healthy plant growth.

   Fertiliser is applied to a spring barley crop throughout the growing season, with 5 tonnes per hectare of poultry manure being applied before the field is ploughed. Nitrogen fertiliser is spread when the crop germinates, at a rate of 2 tonnes per hectare. Following harvest, 7 tonnes per hectare of farmyard manure is spread on the field to enrich the soil, in preparation for the next crop to be grown.

(GO!) Activity 3.1.3 Work as a group to:

1. Each member of the group should have responsibility for researching one principle. This information should then be combined **to make a Powerpoint presentation consisting of 5 slides** to be given to the class. Each member of the group should then speak about their part in the research carried out.

2. Each member of the group is to choose one food domestic crop to research, making sure to include information about the area of the where the crop is grown, its yield, price per tonne, and fertiliser application rates. The website www.ukagriculture.com has plenty of useful information to get you started. All group members' individual information should then be collated into a wall display.

After working on this chapter, I can:

1. Explain that food supply depends on agricultural production, which must be sustainable and not degrade natural resources.

2. State that food security refers to the extent to which food production is assured and sustainable.

3. Explain that increased food production is needed to meet the needs of an increased human population.

4. State that sustainability depends, in part, on using measures that do not degrade the natural resources on which agriculture depends.

5. Describe factors that affect plant growth in terms of cultivar yield, fertilisers, crop protection measures, and competition.

6. State that the area for growing crops is limited.

7. State that livestock produce less food per unit area than plant crops because of the loss of energy between trophic levels.

3.2 Photosynthesis

You should already know:

- The word equation summarising photosynthesis:

$$\text{Carbon dioxide} + \text{water} \xrightarrow[\text{chlorophyll}]{\text{light energy}} \text{glucose} + \text{oxygen}$$

- Photosynthesis occurs in two stages. The first stage is called the light reaction, and the second stage is called carbon fixation.
- In the light reactions, light energy from the sun is trapped by chlorophyll and converted to chemical energy in ATP.
- During the light reaction in the chloroplast, water molecules are split into hydrogen and oxygen.
- Hydrogen from the light reactions attaches to hydrogen acceptor molecules and excess oxygen diffuses out of the cell.
- Carbon dioxide and hydrogen combine during the carbon fixation stage within the chloroplast and use ATP to form sugar.
- Sugar is used as a substrate for cell respiration, for the formation of structural carbohydrates such as cellulose, or stored in the chloroplast as starch.
- Carbon dioxide concentration, light intensity and temperature can affect the rate of photosynthesis and plant growth and are called limiting factors.

Learning intentions

- Explain the role of photosynthetic pigments in the capture of light energy during photolysis.
- Describe the different effects of light energy falling on a leaf.
- Explain the difference between an absorption and action spectrum for the pigments within a green plant.
- State that the pigments chlorophyll a and chlorophyll b absorb most light at the blue and red ends of the visible light spectrum.
- Explain the role of carotenoid pigments within the absorption and action spectrum.
- Describe the absorption of light energy by pigment molecules, and the transfer of high-energy electrons along an electron transport chain to generate ATP.
- State that water molecules are split during photosynthesis to release hydrogen and oxygen.
- Describe the stages of the Calvin cycle (carbon fixation stage), and the importance of ATP and NADPH produced during photolysis.

Process of photosynthesis

Effects of light energy falling on a leaf

Light energy from the sun which falls on a leaf can be absorbed, transmitted, or reflected. Some light energy passes through the plant cells of a leaf, and is absorbed in the chloroplast, to be used in photosynthesis. The remainder of the light energy falling on a leaf is either transmitted or reflected. Transmitted light energy passes all the way through the cells of a leaf, and out the other side. Reflected light energy bounces off the surface of the leaf and is detected by the human eye. Plants look green because all colours (wavelengths) of light are absorbed by the leaf, except green light, which is reflected.

reflected

transmitted

absorbed

Figure 3.2.2 *Fates of light energy falling on a leaf*

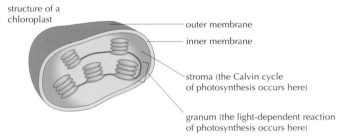

structure of a chloroplast

outer membrane

inner membrane

stroma (the Calvin cycle of photosynthesis occurs here)

granum (the light-dependent reaction of photosynthesis occurs here)

Figure 3.2.1 *Section through chloroplast*

📖 Photolysis

Takes place in the grana and involves the splitting of water molecules into hydrogen and oxygen, using absorbed light energy.

📖 Carotenoids

Pigments xanthophyll and carotene which absorb light in the yellow and orange part of the visible spectrum, and pass energy on to chlorophyll a.

Capture of light energy by photosynthetic pigments

Photosynthetic pigments are found in the chloroplast, which is the location of the light-dependent stages of photosynthesis called **photolysis**. Photophosphorylation also occurs during the light-dependent stage, where light energy is used to phosphorylate ADP to ATP.

Chlorophyll a and the accessory pigment chlorophyll b are the two main pigments in green land plants which absorb most light energy. The **carotenoids** are accessory pigments and pass on their absorbed light energy to chlorophyll a.

Electromagnetic spectrum and visible light

Visible light (which is needed for photosynthesis) is a form of electromagnetic radiation. It travels in waves, and forms part of the electromagnetic spectrum.

Visible light is made up of different colours- or wavelengths- of light. The term wavelength refers to the distance between two peaks of a light wave. The colour varies with the wavelength of light, which is measured in nanometers. One nanometre = 10^{-9} metres.

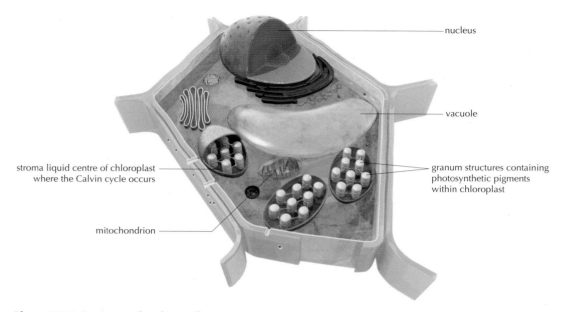

Figure 3.2.3 *Structure of a plant cell*

Photosynthetic pigments absorb different wavelengths of light between the blue (400nm) and red (700nm) ends of the visible spectrum.

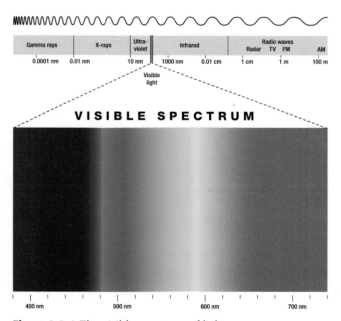

Figure 3.2.4 *The visible spectrum of light*

Absorption spectrum

The wavelengths of light absorbed by individual photosynthetic pigments can be represented as a graph called the *absorption spectrum*.

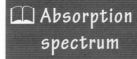

Absorption spectrum

Shows the different wavelengths of light absorbed by photosynthetic pigment.

Figure 3.2.5 shows the **absorption spectrum** for the pigment chlorophyll a, the main light-absorbing pigment in photosynthesis. Peaks occur at the red and blue ends of the spectrum, at 400nm and 700nm wavelength, where absorption of light energy is highest for this pigment. Little light energy is absorbed by chlorophyll a within the yellow and orange part of the spectrum, between 500 and 600nm wavelength, and even less is absorbed in the green region of the spectrum.

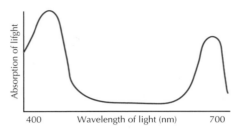

Figure 3.2.5 *The absorption spectrum for chlorophyll a*

Action spectrum

Action refers here to the process of photosynthesis. An **action spectrum** for a green plant shows those wavelengths of light which are best at generating the action of photosynthesis.

Figure 3.2.6 shows peaks at both the blue and red ends of the spectrum, which indicates that these wavelengths of light are best at 'driving' photosynthesis. This is similar to the absorption spectrum of chlorophyll a.

Action spectrum

Shows the rate of photosynthesis at each wavelength of light.

Hint

'Action' in action spectrum refers to photosynthesis.

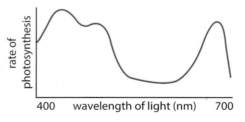

Figure 3.2.6 *Action spectrum for a plant*

The absorption spectrum for a photosynthetic pigment shows the wavelength (colours) of light absorbed by the pigment.

The action spectrum shows the rate of photosynthesis which occurs in a green plant when exposed to different wavelengths of light.

Biochemistry of photosynthesis

Capture of light energy by chlorophyll a

Photosynthetic pigments are found in the chloroplast of a plant cell. Light energy falling on to a leaf is absorbed into the cells and subsequently into the chloroplasts, where it is captured by the photosynthetic pigments.

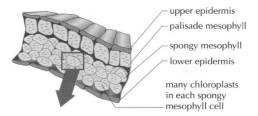

upper epidermis
palisade mesophyll
spongy mesophyll
lower epidermis
many chloroplasts
in each spongy
mesophyll cell

Figure 3.2.7 *Section through leaf*

Photolysis

In the chloroplast, some absorbed light energy is used to break the chemical bonds within a molecule of water, releasing oxygen and hydrogen. Oxygen is released by diffusion from the leaf through the stomata, hydrogen attaches to the **co-enzyme** NADP to form NADPH.

Absorbed light excites electrons in the pigment molecule. These high energy electrons pass along an electron transport chain, generating ATP from ATP synthase.

Some absorbed light energy is used to split a water molecule into hydrogen and oxygen. The hydrogen is picked up by the hydrogen carrier molecule NADP to form NADPH, and the oxygen is released from the leaves through the stomata.

ATP and NADPH produced during the light-dependent stage of photosynthesis are needed for the light-independent stage, called the Calvin cycle.

> **📖 Co-enzyme**
>
> A substance which helps an enzyme to work.

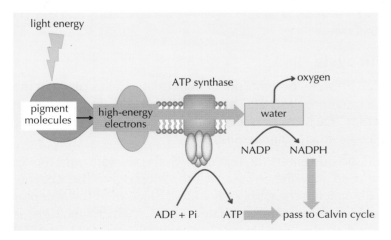

Figure 3.2.8 *Process of Photolysis - Light dependent stage of photosynthesis*

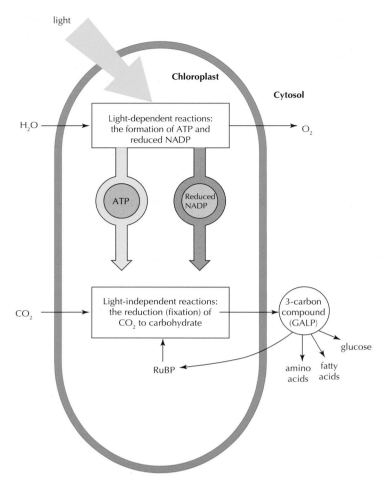

Figure 3.2.9 *Summary of the light dependent and light independent stages of photosynthesis.*

Calvin cycle

The Calvin cycle, or light-independent stage, takes place in the stroma of chloroplasts and results in the production of glucose molecules. Each stage is controlled by enzymes, and so the cycle is sensitive to temperature.

Carbon dioxide enters the leaf through the stomata, and moves by diffusion into the stroma of the chloroplast. The enzyme **RuBisCo** attaches the carbon dioxide molecules to the acceptor molecule ribulose biphosphate (RuBP), which in turn is converted to the compound 3-phosphoglycerate. Using ATP and NADPH molecules produced during the light-dependent stage, 3-phosphoglycerate is converted to glyceraldehyde–3–phosphate (G3P).

Some G3P is used to regenerate RuBP, which is continually being used up as it combines with carbon dioxide, and some G3P is used to produce glucose.

📖 **RuBisCo**

The enzyme which attaches carbon dioxide molecules to the acceptor molecule RuBP.

Some glucose may be stored in the chloroplast as starch grains, and some may be used to manufacture the cellulose needed for new plant cell walls. Glucose molecules may be used as a substrate for cell respiration, or may form new metabolites within other biosynthetic pathways.

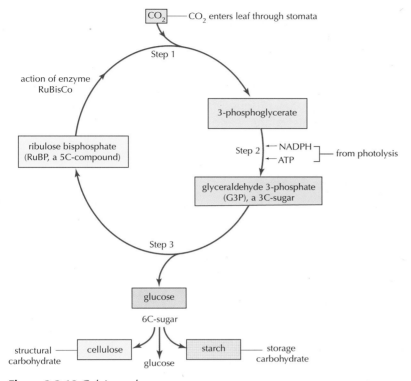

Figure 3.2.10 *Calvin cycle*

🔍 **Hint**

When learning the Calvin cycle, begin where carbon dioxide enters and attaches to RuBP. Try drawing the Calvin cycle out from memory!

The balance of carbon dioxide and oxygen within the Earth's atmosphere is maintained through the process of photosynthesis. Much of the excess carbon dioxide, is absorbed by microscopic photosynthetic phytoplankton within the sea. Burning fossil fuels has resulted in the Earth's oceans being overloaded with carbon dioxide, which the phytoplankton are unable to process. This has resulted in carbon dioxide dissolving in sea water, lowering its pH, and reducing the global marine population of photosynthesising phytoplankton. This will have significant implications for the balance of atmospheric carbon dioxide and oxygen in the future.

GO! Activity 3.2.1 Work individually to:

Restricted response

1. a) State the three outcomes of light energy landing on a leaf.   1

   b) Explain the difference between the terms 'absorption spectrum' and 'action spectrum'. 2

   c) Explain how the conversion of 3-phosphoglycerate to G3P in the Calvin cycle is
dependent on chemical reactions in the light-dependent stage of photosynthesis.   2

Extended response

   d) Describe the role of photosynthetic pigments in the process of photosynthesis.   6

2. a) The following table shows the mass of photosynthetic pigments found in 1cm^3 of the
dry mass of leaves from two different plants, plant X and plant Y.   4

Name of pigment	Mass of photosynthetic pigment (µg/cm^3 of dry mass)	
	Plant X	Plant Y
Chlorophyll a	0·90	0·94
Chlorophyll b	0·32	0·33
Xanthophyll	0·26	0·49
Carotene	0·52	0·67

    i. Calculate the average mass of carotenoid pigments in Plant Y.   1

    ii. Calculate the ratio of carotene to xanthophyll pigment in Plant X.   1

    iii. Calculate the percentage difference of the mass of chlorophyll a in Plant Y,
compared to that of Plant X.   1

    iv. Decide in which plant, X or Y, are the carotenoid pigments likely to absorb more
light energy for photosynthesis and justify your answer.   2

   b) An experiment was set up to measure the rate of photosynthesis in a green plant at
different temperatures by placing the apparatus in a water bath. The concentration of
bicarbonate solution, and light intensity were kept constant each time. The volume of
oxygen produced at each temperature was recorded at 2-minute intervals for 20 minutes.

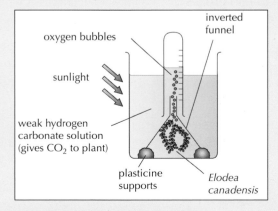

Figure 3.2.11 *Diagram of experiment*

Temperature °C	Time (minutes)	Volume of Oxygen (cm³)
25°C	2	3·1
	4	3·3
	6	3·3
	8	3·4
	10	3·3
	12	3·4
	14	3·5
	16	3·5
	18	3·5
	20	3·5
35°C	2	4·1
	4	4·7
	6	4·8
	8	5·2
	10	5·3
	12	5·4
	14	6·2
	16	6·8
	18	7·3
	20	7·4
45°C	2	0·5
	4	0·8
	6	1·1
	8	1·0
	10	0·7
	12	0·3
	14	0·1
	16	0·0
	18	0·0
	20	0·0

i. Using three different colours of pencil graph paper and a ruler, present this information on a one line graph.   4

ii. State the maximum volumes of oxygen produced at 25°C and 35°C. Give several reasons for this difference.   2

iii. Give a reason why no more oxygen was produced after 14 minutes at 45°C.   1

iv. State the number of minutes over which the volume of oxygen produced remained constant at 25°C.   1

v. Calculate the percentage increase in volume of oxygen produced over 20 minutes at 35°C.   1

Activity 3.2.2 Work in pairs to:

1. With one colour of pencil, each member of the pair is to **draw a labelled diagram** of photolysis and the Calvin cycle from memory, on a piece of A3 paper. Exchange diagrams, and with a second colour of pencil add any further information to the diagram again from memory. This will illustrate your current understanding of the chemistry of photosynthesis. Lastly, work together to add any missing pieces of information to your diagrams using a third colour of pencil, this time referring to your notes and text book. The information written in the third colour represents the work that now needs to be learned.

2. Find the website http://www.biologycorner.com/ then type 'photosynthesis' into the search box. Click on 'photosynthesis simulation'. Use the simulation to find out how wavelength and light intensity affect the rate of photosynthesis. Remember to change only one variable at a time.

Activity 3.2.3 Work as a group to:

Nominate a chairperson, a timekeeper, and scribe. Your teacher will provide each group with five A3 sheets of paper. At the top of each sheet will be written a different word connected with photosynthesis, for example 'chloroplast'. The group should then write down any words connected with that word – e.g., granum, pigments, and so on for five minutes at the most. Each group should then exchange their five word sheets with another group, collected by their chairperson only.

The task: the chairperson will give one sheet, unseen by other members of the group, to each person in turn. Each person must try to describe the word at the top of the sheet to the rest of the group **without** using any of the words written below it!

The correct identification of a word from the description wins one point for the group. The group with the most points wins.

The timekeeper will give each person two minutes to do this, and the scribe will keep the score.

After working on this chapter I can:

1. Name the four photosynthetic pigments in a green plant, and describe their role in the capture of light energy and the process of photolysis.

2. Explain the difference between transmitted, absorbed and reflected light.

3. Explain the role of carotenoid pigments in the absorption of light energy for photosynthesis.

4. Describe the difference between the absorption spectrum and action spectrum for a photosynthetic pigment.

5. State that chlorophyll a and chlorophyll b absorb most light at the blue and red ends of the visible spectrum.

6. Explain the absorption of light energy by chlorophyll, and the transfer of high energy electrons through the electron transport chain on the membrane of the chloroplast.

7. Describe the generation of ATP molecules through the activation of ATP synthase by hydrogen ions.

8. Describe the process of photolysis, the splitting of water molecules to produce NADPH, ATP and oxygen.

9. Describe the stages of the Calvin cycle including the roles of RuBisCo and RuBP.

10. Explain the role of ATP and NADH produced during photolysis for the formation of glyceraldehyde–3–phosphate in the Calvin cycle.

11. State that glucose in a plant cell can be used for cell respiration, synthesis of cellulose for cell walls or stored as starch.

3.3 Plant productivity

Plant productivity

Measurement of plant productivity

📖 **Biomass**

Total mass of living cells.

📖 **Productivity**

Conversion of light energy to chemical energy (carbohydrate), resulting in an increase in biomass due to the assimilation of some glucose into cellular structures.

🔍 **Hint**

Increased biodiversity of producers in an ecosystem results in higher net assimilation because different plant species are able to utilise a wider range of wavelengths of light for photosynthesis.

Some glucose produced in plant cells during photosynthesis is broken down during cell respiration to release energy. For plants to grow, cells must divide and increase in number. This requires energy. The total mass of living plant cells is called the **biomass,** and as a plant grows, this increases.

An increase in abiotic factors such as light intensity, temperature, and carbon dioxide concentration will cause an increase in the rate of photosynthesis. This results in an increase in biomass, or **productivity** of a crop, which can be calculated by measuring the biomass in kilograms per square metre, per year ($kgm^{-2}y^{-1}$).

Productivity of grassland

Productive areas of grassland may be used for grazing cattle or sheep, and contain a broad biodiversity of plant species. Many different species means a wide range of photosynthetic pigments within the grassland ecosystem which will capture a broad range of wavelengths of light. This provides more energy for photosynthesis and the production of carbohydrate. Productive grassland is able to support a larger number of grazing animals than a less productive area of grassland.

Net assimilation

During photosynthesis, some glucose produced in the Calvin cycle is assimilated into cellulose for new plant cell walls. Some glucose is also assimilated into other cell components, such as protoplasm. **Assimilation** of glucose into new cellular structures results in an increase in biomass, and therefore the productivity of the growing plant is also increased.

Some glucose is not assimilated into new cell structures, but instead is broken down in cell respiration to release energy for the cell, in the form of ATP. This results in a decrease in biomass, as glucose is lost from the cell.

Net assimilation = increase in biomass (assimilation) – decrease in biomass (cell respiration)

Net assimilation of a growing plant can be calculated by collecting fresh samples of leaves, and measuring the increase in dry mass per unit of leaf area in grams.

Leaf area index

Leaves are the main photosynthetic structures in plants, and have flat upper surfaces to capture light energy from the sun. Multiple layers of leaves form a tree canopy, and are arranged in a mosaic pattern to maximise light energy capture. This mosaic arrangement reduces shading on the leaves in the lower levels as a result of overlapping, making photosynthesis more efficient.

Figure 3.3.1 *Mosaic arrangement of leaves maximises the capture of light energy and increases the leaf area index*

> 📖 **Assimilation**
>
> Carbohydrate is converted into new cellular structures.

The leaf area index (LAI) of an ecosystem is a measure of the total area of photosynthesising leaves.

It is calculated by measuring the total area of upper leaf surface, compared with the total area of soil. A high leaf area index indicates a high rate of photosynthesis and plant productivity. A low leaf area index indicates low plant productivity.

Leaf area index (m²/m²)	Description
0	Bare soil
1	Single layer of leaves covers entire area of soil
2–6	Multiple layers of leaves in canopy per unit of soil area.

Improving leaf area index of a crop

When a new crop is sown in a field, the seed rate is first calculated to ensure that there is optimum space between seeds as they are placed in the soil by a seed drill. If seeds are too far apart, when the young plants grow light energy will fall between the plants onto the soil and be lost. The crop will have a low leaf area index, of between 0.1 and 0.6.

If the seeds are too close together, the leaves of young plants will overlap, reducing the capture of light energy, and plants will be forced to compete for available resources resulting in a leaf area index of between 2 and 6. If seeds are planted the correct distance apart, all bare soil will be covered with a leaf canopy, and overlap of leaves will be reduced resulting in an optimum leaf area index of the growing crop between 0.8 and 1.0. maximising photosynthetic efficiency.

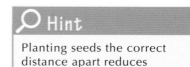

Hint

Planting seeds the correct distance apart reduces intraspecific competition.

Figure 3.3.2 *Tank full of seeds which travel down black hoses into the soil at a predetermined speed as the tractor moves forward, thus ensuring the correct spacing of seeds*

Harvest index

If a field of potatoes is grown and then harvested, the total biomass of plant material (leaves, stems, roots, potato tubers) collected is called the biological yield.

The economic yield, is that part of the biological yield which is desirable and can be sold. In this example, the economic yield is the total mass of potato tubers.

The harvest index is an indication of the economic value of a crop, and is expressed as a percentage value. The higher the harvest index, the more valuable the crop is in terms of economic return.

$$\text{Harvest index} = \frac{\text{dry mass of economic yield}}{\text{dry mass of biological yield}} \times 100$$

Make the link

The ability of food supply to meet consumer demand is discussed in Unit 3.1.

Figure 3.3.3 *Leaves, stems, roots, and potato tubers make up the biological yield; only potato tubers make up the economic yield*

GO! Activity 3.3.1 Work individually to:

Restricted response

Read the following statements and indicate whether they are true or false. If a statement is false, explain why, and rewrite the statement in the correct form.

1. **a)** During plant growth, an increase in productivity usually leads to an increase in biomass. 1

    b) An increase in the concentration of carbon dioxide where light and temperature are not limiting factors, results in a decrease in biomass. 1

    c) The term 'net assimilation' refers to an increase in biomass. 1

    d) A crop with a leaf area index of 4 is a productive crop. 1

    e) If a mature crop is growing in a field which has over 40% of the total area as bare soil, the leaf area index of the crop will be high. 1

    f) Harvest index is an indication of the economic value of a crop. 1

(continued)

Extended response

    g) Discuss how an increase in the atmospheric concentration of carbon dioxide might affect global productivity.   6

2. The following table shows the leaf area and leaf area index for a variety of different crops.

Name of crop	Leaf area (m²)	Leaf area index
Barley	220	3·5
Wheat	200	3·2
Maize	176	2·0
Potatoes	120	1·6
Carrots	100	1·1
Linseed	90	1·0
Oilseed rape	69	0·8
Turnips	60	0·6
Cauliflower	48	0·4
Cabbage	20	0·1

    a) Present each set of data separately in the form of a bar chart. 3 marks for each graph.   6

    b) Calculate the percentage difference in leaf area between potatoes and cauliflower.   1

    c) Describe the relationship between the leaf area and the leaf area index of a crop.   2

GO! 3.3.2 Work in pairs to:

Discuss and analyse the data in the following tables. Write down the different possible conclusions that can be drawn from the data presented referring to (a) the relationship between the height of the grass, and the average leaf area of an individual grass plant and (b) the relationship between height of grass and leaf area index.

Tables show grass height, average leaf area per individual grass plant, and leaf area index of whole field within a 30-hectare grass field on a Scottish Lowland farm, continually grazed by sheep throughout the year.

Month of year	Average leaf area of an individual grass plant (cm²)			
	Height of grass 5cm	Height of grass 10cm	Height of grass 15cm	Height of grass 20cm
December	1·20	2·10	4·32	5·78
January	2·00	1·82	2·23	4·13
May	1·39	1·51	2·46	3·33
August	1·42	3·95	3·57	2·52

Leaf area index of 30-hectare field				
Month of year	Height of grass 5cm	Height of grass 10cm	Height of grass 15cm	Height of grass 20cm
December	2·00	2·68	3·62	3·84
January	1·98	2·59	4·26	4·23
May	1·50	2·56	4·45	4·42
August	2·72	2·91	3·92	3·91

GO! 3.3.3 Work in groups to:

Each member of the group should choose a type of food crop, such as barley, wheat or potatoes for example. Using the information given in this chapter, together with research on the internet, each individual should produce **two powerpoint slides** about factors affecting the productivity of their chosen crop. The group should discuss their findings and show their powerpoint presentation to the class.

After working on this chapter I can:

1. Explain what is meant by the productivity of plants in relation to photosynthesis and biomass.

2. Relate the importance of productivity to global food production.

3. Define the term 'net assimilation' as being the increase in biomass, as a result of light energy being converted into carbohydrate (glucose) minus the mass of glucose broken down during cell respiration.

4. State the link between productivity and leaf area index.

5. Calculate leaf area index by dividing total leaf area by the total area of soil.

6. State that a high leaf area index results in high productivity of a crop.

7. Explain the effect of seed spacing on leaf area index and crop productivity.

8. Explain the difference between biological and economic yields.

9. Calculate a harvest index using biological and economic yield dry mass values.

3.4 Plant and animal breeding

- Explain how the F_2 generation contains a wide variety of genotypes and is a source of genetic variation.
- Describe methods of plant and animal improvement using gene technology.

Improvement of productivity through selective breeding

Selection of desirable genetic characteristics

Within agricultural practice inherited genetic traits which are desirable, in terms of increased productivity of animals and plants, are selected for by farmers and growers to ensure they are conserved in future generations. Only individual animals or plants possessing a desirable genetic characteristic are selected to be included in breeding programmes.

Some examples of desirable genetic characteristics in animals (livestock) and plants (crops) are shown in the table below:

Livestock (e.g. cattle, sheep, pigs, poultry)	Crops (e.g. barley, wheat, potatoes)
Resistance to bacterial infections.	Resistance to pests and diseases.
High conversion of feed to meat (feed conversion ratio).	High conversion of light to carbohydrate (photosynthetic efficiency).
High fertility rate.	Large economic yield.
Ability to thrive in challenging environments, e.g., on Scottish hills or on islands exposed to severe weather.	Ability to grow in heavy/light soils.

By breeding only those organisms which possess a specific desirable characteristic within each subsequent generation, an improved animal breed or plant **cultivar** can be sustained, as can be seen in **Figure 3.4.1**, where Aberdeen Angus cattle have been bred to select characteristics which maximise body size and therefore the mass of saleable meat. **Figures 3.4.2** and **3.4.3** show the differences between strains of wheat, where one has been bred for resistance to disease.

📖 Cultivar

A cultivated variety of plant.

Figure 3.4.1 *Aberdeen Angus cattle have been selected for the desirable genetic characteristics of short legs and a small head, to maximise the body size of the animal*

Figure 3.4.2 *Wheat that possesses the gene for disease resistance*

Figure 3.4.3 *Wheat which lacks the gene for disease resistance*

Field trials

A field trial is an experimental investigation to test the growth of either new cultivars of plants or **genetically modified** plants, in real field conditions. Field trials can also be used to gather data on the effect of treatments, such as different levels of fertiliser application on the growth of plants. The information generated from field trials by research scientists is then passed on to farmers and growers, to help improve the productivity of their crops.

Setting up a field trial

A large field is divided up into rectangular plots of ground. The plots are dug (or cultivated), leaving grass paths between to allow researchers access to the growing plants.

a) New cultivar – if a new improved crop is being tested in field conditions, several plots are sown with seeds and allowed to grow to maturity, then harvested. Information throughout the trial period such as leaf area, dry mass and economic yield are recorded.

b) Genetically modified plants – plots of genetically modified plants are grown alongside plots of the same plant species which has **not** been genetically modified, in order to compare data from each. In this way, any advantage the GM crop has over a traditional variety can be quantified.

c) Testing crop treatments – how a crop responds to treatments such as levels of fertiliser applied, pesticide treatments, or being grown in different soil types, can be measured by comparing a number of replicate 'treatment' plots with a number of replicate 'control' plots which have not received any treatment. The 'treatment' is the only variable changed within the plots; other variables such as soil type, variety of plant and seed spacing are kept constant to ensure a fair test.

Comparative data from plants grown in the two sets of plots can then be analysed.

Randomisation of field trial plots is important to reduce bias in the resulting data. For example, if six neighbouring plots contained a new cultivar of wheat it is possible that plots 1 and 2 are on a slight slope and have a much drier soil than plots 3 to 6. This could influence the results of the trial, causing them to be biased. Improved yield in plots 1 and 2 could be due to the drier soil, rather than the desirable gene being tested within the genome of the plant.

To reduce bias in results, the new cultivar of wheat would be grown in randomly located replicate plots across the whole field trial area.

Figure 3.4.4 *Variety of different field trail plots covering a large area, showing a grass access path*

Breeding animals and plants

Outbreeding

Animals and plants are naturally **outbreeding**. Populations of animals and plants which have been outbred show a wide range of genetic variation. Outbreeding involves crossing individuals of the same species together which are genetically distant from one another. This results in a large number of heterozygous individuals, which are unlikely to express undesirable recessive characteristics in their phenotype due to the presence of dominant alleles in their genotype. Heterozygous individuals are generally stronger and have a good rate of growth. This is known as **hybrid vigour**.

> 📖 **Outbreeding**
>
> Crossing unrelated members of the same species together.

> 📖 **Hybrid vigour**
>
> Describes the desireable genetic characteristics of a heterozygous individual which are superior to both parents.

Inbreeding

Self-pollinating plants are naturally inbreeding, where both male and female gametes are produced by the same flower or the same plant, reducing genetic variation in the resulting seeds.

In farming systems, animals and plants can become inbred through selective breeding programmes. Inbreeding involves crossing closely related individuals together to produce offspring. This can happen if a farmer uses the same ram to mate with a flock of ewes for a number of years. Eventually the ram will mate with some of his own offspring.

Inbreeding depression

Breeding genetically similar individuals together can result in an increase in the frequency of damaging homozygous recessive alleles found in the genotype. This is called **inbreeding depression**, and is a consequence of selective breeding where heterozygous traits are gradually removed over a number of generations when selecting for a desirable genetic trait. An example of inbreeding depression is split eyelids in lambs, which occur as a result of breeding sheep which are genetically closely related.

Inbreeding depression can be avoided by adding new individuals to a flock of sheep or herd of cattle, maintaining genetic diversity while continuing to selectively breed for desirable genetic traits.

Inbreeding depression is reduced in naturally self-pollinating plants, as the process of natural selection removes individuals with damaging recessive alleles from the population.

Monohybrid cross

Production of F_1 hybrids in animals

A cross between two parent animals that differ in the **alleles** they each possess for one specific gene, where one parent has two dominant alleles and the other parent has two recessive alleles, is called a **monohybrid** or **single factor** cross.

All offspring resulting from the cross will possess one dominant and one recessive allele for the gene, and are called **F_1 hybrids**.

If two individuals from the F_1 generation are crossed together, their offspring (the F_2 generation) will have a phenotypic ratio of 3:1 for the dominant and recessive alleles, respectively.

Example: Monohybrid cross in Aberdeen Angus cattle

Figure 3.4.5 *Red coat and black coat phenotypes in Aberdeen Angus cattle.*

Aberdeen Angus cattle have a coat colour gene which has two alleles: black coat and red coat. The black coat allele (B) is dominant to the recessive red coat allele (b).

Parent with black coat × **Parent with red coat**

BB   **bb**

Gametes all **B**   Gametes all **b**

F₁ generation   all **Bb** (all heterozygous for black coat)

If two individuals from the F$_1$ are crossed together then:

Parent with black coat × Parent with black coat

Bb   **Bb**

Gametes 50% = **B**, 50% = **b**   Gametes 50% = **B,** 50% = **b**

The table shows what happens when the gametes are placed in a Punnett square to find possible combinations of gametes at fertilisation.

Gametes	B	b
B	BB	Bb
b	Bb	bb

Phenotypic ratio of offspring = 3 black coats to 1 red coat.

Improved F$_1$ hybrids – animals

Crossing different breeds of sheep together promotes outbreeding, increased genetic variation, **hybrid vigour**, and the production of lambs with a wide range of genetic characteristics.

> ### 🔍 Hint
> Remember, individuals are homozygous when alleles in the genotype are the same, for example BB, or bb. Individuals are heterozygous when alleles in both genotype are different, for example Bb.

> ### 📖 Hybrid vigour
> Offspring from a cross of genetically different parents show increased strength and resistance to disease.

Figure 3.4.6 *Scottish Mule breed – thick fleece and larger body*

The Scottish Blackface sheep is a hardy breed, able to withstand severe weather on exposed Scottish hills. This is due to the sheep having a thick, long fleece. Their lambs, however, are small, and take a long time to reach maturity.

The Bluefaced Leicester sheep is a Lowland breed, with short, fine wool and large, early maturing lambs.

Crossing these two breeds of sheep together can produce larger lambs with thick fleeces able to thrive on exposed hill pasture. These lambs are improved F_1 hybrids and are known as a crossbreed called the Scottish Mule sheep.

Back cross

The Scottish mule is an F_1 hybrid which has a thick fleece. Individual sheep in the F_1 generation may be heterozygous or homozygous for thick fleece. A farmer needs to keep homozygous individuals for breeding in order to maintain the desirable characteristic of a thick fleece. The farmer can find out if potential breeding animals are heterozygous or homozygous for thick fleece by carrying out a **back cross**.

A back cross involves crossing the individual with the unknown genotype for thick fleece, **F**, (heterozygous or homozygous), with another which is a known double recessive for fine fleece, **f**. The answer is found by examining the phenotypes of the offspring.

Thick fleece = **FF or Ff?**

Fine fleece = **ff**

Is the sheep **FF** or **Ff** for thick fleece?

back cross 1

Parent = **FF or ff** × Parent = **ff**

The F_1 result all have thick fleece, therefore must have one dominant allele: **Ff**.

To produce this result in the F_1 generation, the parent with the unknown genotype must be homozygous: **FF**.

back cross 2

Parent = **FF**   ×   Parent = **ff**

F_1 result   50% offspring have thick fleece and 50% have fine fleece

To produce this result in the F_1, the parent with the unknown genotype must be heterozygous: **Ff**.

📖 Back cross

A back cross is used to determine whether a potential animal or plant 'parent' is homozygous (true breeding) or heterozygous for a desired trait. By examining the phenotypes of the resulting offspring in the F_1 generation, the genotypes of the unknown parent can be determined.

🔍 Hint

In a back cross, the unknown genotype individual is always crossed with a homozygous recessive individual. The answer to the possible genotype of the unknown individual is found by examining the F_1 offspring.

Following the test cross the farmer would select the known homozygous parent for thick fleece to breed from, maintaining the desirable thick fleece characteristic within the flock.

The Scottish Mule breed, with desirable characteristics of thick fleece and larger lambs, can also be maintained by crossing Scottish Blackface ewes with Bluefaced Leicster rams each year, producing improved F_1 hybrid lambs.

Improved F_1 hybrids – plants

Plants that are used as a food crop are often inbred to maximise yield. Courgettes are a vegetable crop with many different cultivars, produced as a result of inbreeding to maximise the size of the fruit, which is the saleable part of the plant.

If two different cultivars of courgette are crossed together, the result is an F_1 hybrid plant which produces a larger fruit, with increased resistance to fungal disease. This is an example of hybrid vigour in plants.

Figure 3.4.7 *Yellow courgettes with resistance to fungal disease*

If plants from the F_1 generation are crossed together, the result is a great deal of genetic variation within the F_2, such as differences in fruit size, colour and disease resistance. The beneficial hybrid vigour within the F_1 becomes 'diluted' and less useful to plant breeders in maintaining the desirable characteristics of the F_1. However, genetic variation within the F_2 is a potential source of new genetic varieties.

A test cross can be carried out with individual plants in the F_1 to identify those plants which have unwanted heterozygous recessive traits. These plants are rejected for breeding, removing unwanted recessive alleles, to prevent them appearing in the next generation.

Figure 3.4.8 *F1 Hybrid courgette with large fruit and fungal disease resistance*

Animal and plant improvement using gene technology

Genome sequencing

Using the polymerase chain reaction technique and bioinformatics, the entire genome of an organism can be analysed, and each gene present within the DNA of an organism can be identified. Base sequencing is then carried out to produce a genome sequence, which can be analysed to identify the presence or absence of a desirable allele of a gene. If the required allele is present in the genome, the organism will be selected and used within a future breeding programme, to ensure the desirable characteristic is passed on to subsequent generations of either animals or plants.

Genetic transformation

Using the process of genetic engineering, a desirable gene from one species of plant or animal can be inserted into the DNA of a different species. The resulting organisms which possess the inserted gene are referred to as being transgenic. The required gene is identified on the DNA of the donor organism, removed and inserted into the plasmid of a bacterial cell or the DNA of a virus, which act as a vectors. The vector transports the required gene into the host organism.

Transgenic organisms which carry the inserted desirable gene on their DNA are then used in breeding programmes, where the inserted gene is passed on to the next generation.

Figure 3.4.9 *Tomato seedlings being grown in petri dishes*

Make the link

Genome sequencing and the process of PCR is described in detail in Unit 1. Vertical and horizontal gene transfer result in genetic variation within a species, and are also explained in Unit 1.

Transgenic animals

A flock of Dutch milk sheep have had the human gene for the production of a protein called Factor 8 which clots blood inserted into their DNA. These sheep excrete Factor 8 in their milk, which is then extracted and used to treat the human disease Haemophilia. Sufferers of the disease lack a protein needed to clot blood, and are in danger of excessive bleeding from even a small cut in the skin.

Hint

In the United Kingdom, sheep are only able to breed once per year in the autumn. Following a gestation period of 147 days, lambs are born in the spring at a time when there is plenty of grass to eat. An important part of the farming year involves bringing sheep from the fields into handling pens, where each individual animal is assessed for future breeding potential. Checks are carried out on the body condition of an animal, how much milk it is likely to produce, and whether the condition of the teeth will allow efficient feeding during the winter. Weak individuals are removed from the flock and sold, often for meat. Sheep which have been selected for desirable traits then run with a ram during the months of October to December, giving birth to strong lambs during the months of March and April the following spring.

Transgenic plants

For people in Asia and Africa, rice is the main source of food. If people have a poor diet and are unable to supplement rice with fresh fruit and vegetables, they will lack vitamin A which is needed for a strong immune system and healthy eyesight. A modified cultivar of rice has now been genetically engineered which is able to synthesise B-carotene, a source of vitamin A. This cultivar is called Golden Rice, and has been produced by inserting a gene from maize and a gene from a bacterium into the DNA of ordinary rice. The ability to synthesise B-carotene gives the rice a golden colour compared to ordinary white rice.

Make the link

The process of genetic modification was discussed in Unit 2.

GO! Activity 3.4.1 Work individually to:

Restricted response

1. Give examples of **two** desirable genetic characteristics in both animals and plants. 4
2. Give the term used to describe the breeding together of genetically similar organisms, increasing the frequency of recessive alleles. 1
3. **a)** A cow, homozygous dominant for black coat, was crossed with a bull homozygous recessive for white coat. Describe both the phenotype and genotype of the calves in the F_1 generation. 2
   b) Two individuals from the F_1 generation were then crossed together. Give the possible genotypes and phenotypic ratio of the F_2 generation of calves. 2

Extended response

Describe how the growth of a new cultivar of barley in local weather conditions can be assessed using a field trial. 6

GO! Activity 3.4.2 Work in pairs to:

Read the following paragraph and use the information to **construct a plan** of the field trial site, on a sheet of A3 paper (not drawn to scale). Each plot should be labelled with the name of the grass species grown, replicate number and level of fertiliser applied. (Begin by working out the total number of plots within the field.)

A field trial was set up within a grass field measuring 1200m by 3245m, to investigate the effect of different levels of nitrogen fertiliser application on the growth of four different species of grass.

Fertiliser rates of 0·5kg and 1·0kg were applied to 3m by 6m plots of the first grass species, *Annual meadowgrass*. Each plot was replicated three times, and situated at random locations across the field area.

This procedure was repeated for the remaining grass species, *Timothy, Crested dog's tail,* and *Yorkshire fog.*

GO! Activity 3.4.3 Work in groups to:

Have a whole-class debate on the advantages and disadvantages of genetically modified animals and plants as a source of food. Divide the class into four groups, to research and present a soundly justified argument on one of the following aspects of GM animals and plants:

Group 1 – Argue **for** GM animals as a food source for humans

Group 2 – Argue **against** GM animals as a food source for humans

Group 3 – Argue **for** GM crops as a food source for humans

Group 4 – Argue **against** GM crops as a food source for humans

The presentations from each group should be judged using the following criteria:

* Number of relevant points made in argument
* Use of examples to illustrate points made

(continued)

- Argument is justified with supporting evidence
- Contributions made from each group member
- Clarity of speech
- Use of visual aids

Useful websites

GM animals

www.bbc.co.uk/ethics/animals/using/biotechnology_1.shtml

www.genewatch.org/sub-572167

GM crops

www.gmo-compass.org/eng/grocery_shopping/crops

www.telegraph.co.uk/news/earth/agriculture/geneticmodification/

After working on this chapter I can:

1. State examples of desirable genetic characteristics, which can be selected for in order to improve production of animals and plants.

2. Describe how a field trial is used to test different aspects of crop growth in real field conditions.

3. Explain the contribution of outbreeding to genetic diversity, hybrid vigour and the generation of improved F_1 hybrids.

4. Explain the advantages and disadvantages of inbreeding, and the effects of inbreeding depression in animal production systems.

5. Describe how a back cross may be used to maintain desirable alleles within a new breed.

6. State that a test cross may be used to identify plants with undesirable heterozygous recessive traits.

7. State that the F_2 generation contains a wide variety of genotypes which are a source of genetic variation.

8. Describe the inheritance pattern of a gene by using a monohybrid cross diagram.

9. Describe how the process of genetic engineering has been used to produce transgenic animals and plants, giving one example of each.

3.5 Crop protection and livestock welfare

You should already know:

- The effects of intraspecific and interspecific competition on organisms within an ecosystem.
- Natural predators can be used to control pest populations. This is a method of biological pest control.
- Biological control and genetically modified crops are alternatives to using pesticides and fertilisers.

Learning intentions

- Describe the impact of weed growth on the productivity of a crop in terms of competition.
- Describe the properties of both annual and perennial weeds, with regard to lifecycles and physical adaptations to reduce competition.
- Explain how damage to crops is caused by invertebrate pest species, and crop diseases are caused by microorganisms.
- State the ways in which weeds, pests and diseases may be effectively controlled, explaining the advantages and disadvantages of each control method.
- Explain the principles of biological control and integrated pest management as methods of pest control in crops, together with their limitations.
- Describe the costs, benefits and ethics of animal welfare in livestock production systems.
- Describe examples of specific animal behaviour which are indicators of stress.
- Explain how animal behaviour can be recorded using an ethogram.
- Describe the use of preference tests and motivation as indicators of animal welfare.

Crop protection

Weeds

Productivity of crops is affected by **competition** with weeds. A 'weed' is defined as simply a plant growing in the wrong place. A field planted with rows of potatoes may be contaminated with weeds growing both between the rows of potato plants, and between the individual plants themselves.

Figure 3.5.1 *Rows of recently sprouted potatoes growing in field*

Weeds compete with growing crops for available light, water, CO_2 and soil nutrients. This is an example of interspecific competition between different plant species, the weed plant and the crop plant.

Common weed species which may contaminate a growing crop are:

- Couch grass
- Perennial ryegrass
- Nettles
- Creeping buttercup
- Plantain
- Horsetail
- Ragwort
- Chickweed
- Daisy
- Goosegrass
- Knotweed
- Shepherd's purse

Properties of annual and perennial weeds

An annual weed has a lifecycle of only one year, and then will die. A perennial weed has a lifecycle of many years, and will continue to grow in a particular place year after year. Each type of weed has characteristic adaptations as shown in the following table:

Figure 3.5.2 *Creeping buttercup*

Comparison of annual and perennial weeds

Annual weeds	Perennial weeds
• Sexual reproduction	• Asexual reproduction
• Short lifecycle (1 year)	• Long lifecycle (2 years +)
• Rapid growth	• Broken pieces of plant can root, and grow into new individual plants
• High numbers of seeds produced	
• Dormant seeds remain viable for a long period of time	• Storage organs provide food for plant in autumn/winter
• Examples: goosegrass, knotweed, shepherd's purse	• Examples: couch grass, nettles, buttercup

Control of weed growth

Both the biological and economic yield of a crop can be reduced significantly by competition from weeds. Therefore it is imperative that weed growth is controlled. This can be done by:

1. Cultural methods
2. Chemical and biological control of pests

Cultural methods

1 Ploughing and weeding

Weeds can be physically removed from a growing crop by employing people to dig out individual weed plants, using hand hoes. This is costly in terms of paid labour.

One man can drive a tractor pulling a small plough, or inter-row cultivator, turning over the soil between rows of crop plants, breaking the roots and above-ground parts of weed plants. This type of cultivation method must be repeated regularly to prevent weeds returning due to the germination of weed seeds which remain in the soil, in order to keep the soil clear between the rows of growing crop plants.

This method is less labour intensive and therefore less costly, particularly if machinery is shared between farmers.

2 Crop rotation

Growing four different crops in one field over a four-year cycle is called **crop rotation**. This practice helps to break the lifecycle of pest species specific to each of the crops used in the rotation. It also prevents depletion of the minerals within the soil, as each crop has different nutrient requirements.

Crop rotation cycle (20-hectare field)

Year 1 – Barley

Year 2 – Potatoes

Year 3 – Turnips

Year 4 – Grass

At the end of year 4, the field would be planted with barley at the beginning of the next crop rotation cycle.

Chemical and biological control of pests

1 Chemical pesticides

> ### 📖 Herbicide
> A chemical which kills plants.

A **herbicide** is a chemical which kills weeds. They are applied in solution as a fine spray. Herbicides are classified according to their action, which may be **contact**, **selective** or **systemic**.

Contact	Selective	Systemic
Non-selective, kills all green plants on contact.	Attacks only broad leaved plants. Narrow-leaved plants such as grass are unaffected.	Herbicide absorbed into transport system of plant (xylem and phloem). Kills all parts of plant.

Figure 3.5.3 *Spraying selective herbicide between rows of growing crops*

Contact herbicides are absorbed through the surfaces of leaves and stems of all plants, and cannot be applied to a growing crop as a spray. Contact herbicides may be used to clear a field of weeds before planting a crop, or by careful application between the rows of a growing crop using a finely directed spray.

Selective herbicides can be sprayed onto a growing crop of cereal plants which have narrow leaves, and remain unaffected. Weeds are mainly broad-leaved plants which have a large surface area for the absorption of a chemical herbicide, which often contains a plant growth hormone. The more chemical that is absorbed, the greater the rate of cell division, using up the plant's reserves of energy and resulting in cell death. Narrow-leaved plants such as wheat and barley absorb a little chemical herbicide due to the small surface area of their leaves, but not enough to kill the plant.

Systemic herbicides are absorbed through the leaves of all plants into the xylem and phloem vessels, to be transported to all cells, including those in the root system, which are killed. This prevents any regrowth of the weed species.

Systemic herbicides cannot be used on a growing crop because the crop plants would be killed, but may be used to clear a field of weeds before soil cultivation and planting of a crop.

2 Control of pests
Crop pests are mainly invertebrate animals, which physically damage all or parts of a growing plant. They include molluscs, insects and nematode worms.

Molluscs and insects feed on leaves, causing damage and reducing photosynthetic efficiency. Nematode (round)worms live in soil and cause damage to the roots of growing plants, reducing the uptake of water and minerals.

Crop pests

Figure 3.5.4 *Snails can eat the leaves of a growing crop*

Pest	Examples	Effect
Insects	greenfly, blackfly	causes leaf damage
Nematodes	roundworm	attacks roots and storage organs
Molluscs	slug, snail	causes leaf damage

Pesticides

A **pesticide** refers to a chemical which kills pests on the leaves of a crop when applied as a fine spray. Contact pesticides kill pests that are present on leaves at the time of spraying. Pesticides are usually specific to one type of pest species, so it is possible that one crop may have to be sprayed with several different chemicals in order to reduce the populations of different pest species. The crop can become infected with pests once again, following the initial spraying.

Systemic pesticides are absorbed by the leaves and enter the xylem and phloem vessels of the crop plant. This causes no harm to the plant, but when insects pierce the phloem vessels with their mouthparts to feed on glucose, they take in the pesticide and are killed.

> 📖 **Pesticide**
>
> A chemical which kills invertebrate animals.

> 🔍 **Hint**
>
> Remember, **systemic** pesticides enter the transport **system** of a plant.

Control of plant diseases

Bacteria, fungi and viruses are microorganisms which can damage the leaves of plants, affecting photosynthesis. They can reach a crop through contaminated seeds, droplets of moisture in the air carried by wind, and being carried by insects.

Diseases

Organism	Example	Disease caused
Fungi	yellow rust on leaves of cereal plants; brown rot of stone fruits, peaches and plums	
Bacteria	affects stems, roots and leaves	leaf spots, blights, galls
Virus	yellow mosaic on lettuce leaves, reduces photosynthesis	lettuce mosaic virus

Fungicides are chemicals which are applied to growing crops to prevent fungal attack.

Figure 3.5.5 *Strawberry leaf with fungal disease*

Fungicide

A chemical which kills fungal spores.

Contact fungicides

A contact **fungicide** is sprayed directly onto the leaves of a growing crop, preventing the germination of fungal spores which are carried by the wind. Spray treatment must be repeated several times during one growing season, due to the fungicide being washed off by rain, as well as the continual growth of new untreated leaves.

Systemic fungicides

Systemic fungicides are sprayed onto the leaves of a growing crop, and the chemical is absorbed into the xylem and phloem vessels of each plant. This prevents the establishment and growth of fungal hyphae, which invade leaf cells.

Figure 3.5.6 *Botrytis cinerea, a fungal infection of tomatoes*

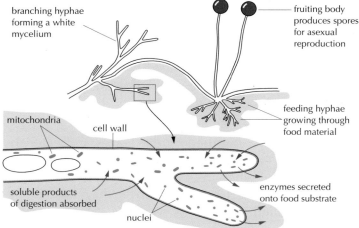

Figure 3.5.7 *Growth of fungus on a leaf*

Air humidity and temperature can be useful indicators of the spread of fungal spores within a specific geographical area, and the potential subsequent need for preventative crop spraying with a fungicide. Protection of a crop against a fungal attack is more effective than treating a crop that has already been infected.

Disadvantages of chemical control methods

Some chemicals contained in pesticides, herbicides and fungicides do not break down over time, and are described as being persistent within the environment.

Persistent chemicals may be washed down into the soil by rain, killing harmless soil organisms and affecting the balance of food webs. These chemicals may enter a food chain and accumulate in the body tissues of animals.

DDT (Dichlorodiphenyltrichloroethane) is a pesticide used in the 1950s to kill mosquitoes carrying malaria. During this time DDT, which is persistent, entered the soil, rivers and lakes. It then entered an aquatic food chain, taken in by **zooplankton** feeding on contaminated **phytoplankton**. The zooplankton were eaten by fish, which in turn were eaten by birds of prey. At each stage of the food chain, the concentration of DDT increased in the body tissues of each organism. This is known as **bioamplification**. The highest concentration of DDT occurred in the muscle tissues of the bird of prey, who laid eggs with extremely thin shells as a result. These eggs broke easily and the young did not survive. By 1980, the USA had banned the use and export of DDT as a pesticide.

📖 **Zooplankton**

Single-celled animals which live in water, and feed on phytoplankton.

📖 **Phytoplankton**

Single-celled plants which live in water.

📖 **Bioamplification**

Concentration of DDT in animal tissues increases at each trophic level of a food chain, from producer to predator.

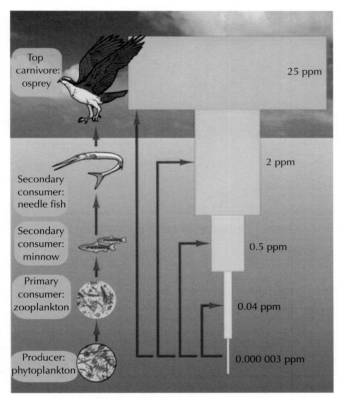

Top carnivore: osprey — 25 ppm

Secondary consumer: needle fish — 2 ppm

Secondary consumer: minnow — 0.5 ppm

Primary consumer: zooplankton — 0.04 ppm

Producer: phytoplankton — 0.000 003 ppm

Figure 3.5.8 *Accumulation of DDT through aquatic food chain (ppm = parts per million)*

Resistance to chemicals

Insect pests may become resistant to chemicals, including DDT, through mutation and natural selection. Frequent exposure to a chemical greatly increases the chance of a resistant pest species emerging.

Biological pest control

Biological methods of pest control reduce the size of the pest population, but they do not eliminate it completely. Used as an

Make the link

Pests have a parasitic relationship with crop plants, where the pest benefits in terms of food but the plant is harmed in terms of cellular damage. Parasitic relationships are described in more detail in the next chapter, on symbiosis.

Make the link

Productivity of a crop is dependent on photosynthetic efficiency, and was discussed in Unit 3.3

Hint

Biological pest control depends upon the relationship between predator and prey.

alternative to chemical control methods, the impact of chemicals on the environment is greatly reduced.

The population size of a pest species is controlled by the introduction of a predator species. The pest species is the primary food source for the predator, resulting in a decrease in the size of the pest population.

Tomato plants grown in greenhouses are often affected by whitefly, which damage the leaves, and so reduce photosynthetic efficiency. The beetle *Delphastus pusillus* has been introduced into greenhouses to control whitefly on tomato plants, as it feeds on both adult flies and larvae.

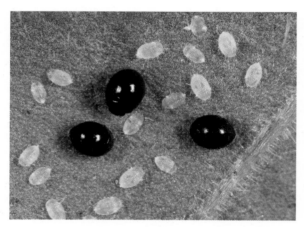

Figure 3.5.9 *Predator beetle* Delphastus pusillus *feeding on whitefly*

Methods of biological pest control have less of a negative impact on the environment than chemicals. The disadvantage, however, is that the predator population may increase so much that it also becomes a pest, upsetting the balance of populations within the food web to which it belongs.

Integrated pest management (IPM)

This system uses a combination of pest-control methods - cultural, chemical and biological - to control the population size of a particular pest. Information on the lifecycle of the pest species is collected, and then an economic approach to pest control is developed which carries the least risk to people and the environment.

1. Pest numbers within a crop are monitored, and a decision is made after identifying the population density, beyond which the pest becomes a significant threat to the economic yield of the crop.

2. Many species of invertebrate may be found within a crop, so any species present which could cause immediate harm is identified as the main pest.

3. Appropriate combinations of measures are then adopted. Chemicals are used sparingly, with only infected plants being sprayed, rather than the whole crop.

As concern grows regarding the impact of chemicals on the environment, integrated pest management is increasingly becoming the method of choice amongst farmers and growers.

Animal welfare

Animals farmed as a source of food are usually kept in artificial conditions in order to maximise economic yield of saleable food, e.g., meat, milk, or eggs.

Summary of main animal production systems within UK agriculture

Production system	Animal management	Animal product
Poultry (hens)	1. Housed free on barn floor. 2. Free-range outside in fields. 3. Battery wire cages (10 hens per 0.25m² cage)	Eggs White meat (chicken)
Sheep	1. Hill grazing throughout year. 2. Lowland grazing, housed in winter, fed on hay/silage/concentrates.*	Lamb meat (male lambs up to one year old), best female lambs kept for breeding. Mutton (meat from sheep older than one year).
Cattle 1. Dairy cattle	Grazing outside during spring and summer, housed in winter, fed on hay/silage/concentrates.	Milk Male calves to be reared for meat. Female calves enter herd as replacements for older cows.
2. Beef cattle		Meat (red) – steak, mince, stew, burgers. Male calves to rear for meat (to 18 months old), female calves kept for breeding.
3. Suckler cows		Produce male calves to rear for meat (to 18 months old). Female calves kept for breeding, enter herd.
Pigs	1. Free-range outside in fields containing shelters. 2. Battery system, pigs kept in metal cages throughout year.	Pork meat – bacon, pork chops, pork sausages, pork joints. Male piglets fattened for meat. Female piglets kept for breeding.

* Hay – grass cut and dried in summer

  Silage – grass cut, baled and wrapped in plastic; the grass ferments in anaerobic conditions, preserving feeding value

  Concentrates – dried pellets of grass containing added vitamins and minerals

Intensive animal production systems

Figure 3.5.10 *Intensive pig farming system*

Intensive or 'battery' systems reduce inputs such as feed, due to the restriction of animal movement. Labour costs are also reduced, as the feeding of animals is often mechanised. With lower financial inputs, economic yield is more profitable. This may occur at a cost to animal welfare.

However, high levels of animal stress reduce productivity and may cause reproductive failure. It has now been recognised that it is in the interest of the farmer to invest in animal welfare and maximise productivity. As a result, within the agricultural industry there is now a move away from intensive animal production systems, toward more extensive systems which provide animals with more natural living conditions.

Recognising stress in farm animals

Poor welfare conditions, such as animals being confined in small cages, result in high levels of **stress**, leading to the demonstration of specific behaviours which may be observed. A 'behaviour' is the observable response of an animal to an internal or external stimulus.

📖 Stress

The total response of an animal to environmental demands.

Altered activity

Animals kept in poor conditions will become less active as a result of stress. Caged animals spend a great deal of time sleeping in order to avoid the stress of their living conditions. The opposite is also true. Animals experiencing high levels of stress can become **hyper-aggressive** and dangerous to humans. Caged animals may bite or charge if approached.

Stereotypy

Animals kept in conditions where they do not have enough room to move may engage in a repetitive, purposeless behaviour. Stereotypic behaviour of this type is seen in pigs confined in metal crates where they cannot turn around. They begin to continuously chew the bars of the crate.

Lions and tigers kept in small areas within a zoo may be observed continuously pacing up and down, which another example of stereotypy.

Figure 3.5.11 *Tiger pacing in a cage – an example of 'sterotypy'*

Misdirected behaviour

Battery hens crammed into a small metal cage experience high levels of **stress**, observed through misdirected behaviours. These include birds pulling out their own feathers, or pecking at their own body, causing injury. Misdirected behaviour is targeted towards the animal causing itself harm.

Lower reproduction levels

Animals experiencing high levels of stress avoid social interactions with others, which can interrupt natural reproductive behaviour patterns. In cases of extreme stress, animals will not produce gametes, or even reabsorb a growing embryo.

Figure 3.5.12 *Intensively farmed chickens for egg production*

Stress may also result in the breaking of the bond between parent and offspring. During lambing, if ewes are chased by a dog, they will abandon their lambs and refuse to let them feed again when reunited.

Ethology – the observation of animal behaviour

In order to study the behaviour of animals in certain conditions, an **ethogram** is first constructed, which lists all the different behaviours shown by the animals being observed, within a specific period of time.

For example, a researcher might observe a flock of 60 free-range chickens in a field for one hour, recording all the different behaviours seen within the flock. The researcher must remain hidden, so that their presence does not affect the behaviour of the chickens. (This is known as the Hawthorne effect, which could affect the validity of the observations.)

Figure 3.5.13 *Free-range chickens*

📖 Ethogram

A list of the range of behaviours observed in an animal.

Ethogram for free-range chickens

- Pecking ground
- Walking
- Stretching wings
- Grooming feathers
- Dust bathing
- Pulling up blades of grass
- Drinking
- Scratching ground with feet
- Clucking
- Short flights
- Standing still
- Perching on low branch
- Chasing other chickens

🔍 Hint

Animal behaviours can be measured by recording:

Duration – length of time the behaviour lasts;

Frequency – number of times a behaviour occurs within a period of time;

Latency – time between a stimulus being applied and a change in behaviour.

Figure 3.5.14 *Chickens in a barn*

If these chickens were then housed in a barn, a second ethogram would be constructed and any differences in behaviours noted. This would give an indication of the effect of changing the environment on the behaviour of the chickens, which could indicate stress. The environment in the barn could then be altered to reduce stress, observed by the return of behaviours recorded when the chickens were in the field.

When the ethogram from the field matched closely the ethogram from the barn, then the conditions in the barn would be in an ideal state for the chickens.

Using a preference test

The ideal environment for housed animals can be determined by using preference tests, where animals are given a choice by indicating a preference for a specific condition.

For example, the floor of a large cattle shed is marked off into four large sections. A different type of floor covering is introduced into each section.

Section 1 – sawdust	Section 3 – shredded paper
Section 2 – straw	Section 4 – dried peat

Hint

If animals are given a choice, stress is reduced.

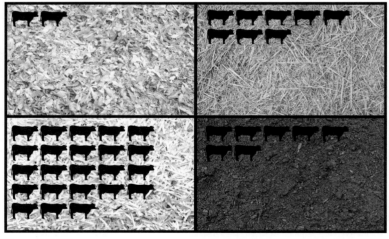

Figure 3.5.15

A herd of 40 beef cattle are released into the shed and observed. After walking around the shed and smelling the different floor coverings, the cattle settle and begin to lie down. Researchers observed the following results:

Section 1 – sawdust	2 animals
Section 2 – straw	8 animals
Section 3 – shredded paper	23 animals
Section 4 – dried peat	7 animals

The preference test results showed that the cattle preferred to lie down on the shredded paper, indicating that this would be the best floor covering to use in the cattle shed in order to provide optimum living conditions and reduce stress.

Motivation in animals

Animals need to be free from discomfort and have access to food, water, air and a comfortable environment in which to sleep. Motivation drives animals to try to satisfy each of these needs when required. If an animal is thirsty, it will be motivated to try to find water. A hungry animal will be motivated to find food. If all needs are satisfied, an animal will lack motivation.

Motivation can be measured quantitatively through an experimental approach.

For example, if some feed is placed in a trough at one end of a barn and an individual sheep is released at the other end of the barn, the time it takes the sheep to find the food can be recorded. The shorter the time taken to find the food, the stronger the motivation.

Motivation can also be measured using a preference test approach. If a sheep is given a choice of food or a bed of deep, clean straw and it moves into the straw, then it was more highly motivated to access clean bedding than to eat food.

🔍 Hint

Consumers are becoming more aware of food production systems, and increasingly select meat which has been ethically reared in humane conditions.

Every animal born on farms in the UK is registered and identified by an ear tag, which means that any meat can be traced back to the source farm. It is in the interest of the farmer therefore, to ensure that animals coming from their farm have been ethically reared, and are disease-free.

Many consumers have begun to select free-range eggs over those produced by battery hens, preferring to support extensive, rather than intensive, production systems.

GO! Activity 3.5.1 Work individually to:

Restricted response

1. State **three** differences between annual and perennial weeds.   3
2. Describe how the growth of weeds can be controlled using cultivation methods.   2
3. Explain the difference between a contact herbicide and a systemic herbicide.   1
4. Identify the three main groups of crop pest.   3
5. The following table shows the number of tonnes of pig meat (pork, bacon and ham) imported into the UK from other EU countries in 2013 and 2014.

United Kingdom Customs Import Statistics		
UK pig meat imports		
Tonnes	**2013 October**	**2014 October**
Total pig meat (inc. offal)	79,514	76,461
Pork	31,882	29,857
From EU	31,762	29,779
Denmark	7,793	7,634
Germany	7,071	6,202
Netherlands	4,909	4,823
Ireland	2,358	2,513
Belgium	2,978	3,157
Bacon/ham	24,119	23,674
From EU	24,119	23,674
Denmark	9,999	10,342
Netherlands	8,847	7,517

(Source – www.bpex.org.uk/prices-facts-figures)

a) i. Calculate the percentage decrease in total pig meat imports between October 2013 and October 2014.   2

    ii. Calculate the percentage difference in tonnes of **pork** imported between 2013 and 2014 for Denmark, Germany, Netherlands, Ireland and Belgium. Present this information as a bar graph.   5

b) The following diagram shows a crop rotation plan for four fields in year 1. Draw three similar diagrams to show the position of each crop in years 2, 3, and 4.   3

Field1 Potatoes	Field2 Grass
Field3 Barley	Field4 Turnips

Extended response

Discuss the advantages and disadvantages of biological pest control.   6

Activity 3.5.2 Work in pairs to:

Restricted response

1. **Construct an ethogram** for a group of horses in a field by copying the following table, and viewing the video 'Horses in the field (stabilised)' on YouTube. Alternatively, you could observe the behaviour of any other animal in school or at home. Then construct an ethogram for the animal, listing all the behaviours observed and how frequently they occurred within a chosen period of time.

Ethogram:

Observed activity	Definition	Frequency (tally) in 30-minute observation period
For example: 1. Tail swish	Tail moved left to right, or up and down.	ⅢⅢ ‖ (7)

a) What was the total number of different behaviours observed over a chosen period of time?

(continued)

b) Which behaviour occurred most frequently? Can you give a possible explanation for this result?

c) Present your frequency and activity data as a bar graph.

2. From the ethogram, choose two interesting behaviours. These are called 'events' as they are short, specific movements which occur during the period of time the animal is observed. Using a stopwatch, observe the video or animal again, and time how long each chosen behaviour (or event) lasted each time it occurred. Record the results in a table. This is called the *duration* of an event.

a) Calculate the average duration time of each chosen event.

b) Which behaviour had the longest duration time?

c) Think of a possible reason why the animal might engage in this behaviour for a longer period of time.

GO! Activity 3.5.3 Work in groups to:

1. **Create a wall display** presenting information about integrated pest management, which is a system of controlling pests and diseases in crops. The wall display should include information on the advantages and disadvantages of IPM as a system of pest control. Information on integrated pest management may be found on the Scottish Natural Heritage website.

2. **Produce a 5 slide Powerpoint presentation** on the guidelines for the welfare of hens, pigs or beef animals in farming systems. For information, go to the website of the Farm Animal Welfare Advisory Council at www.fawac.ie

After working on this chapter I can:

1. Explain how the growth of weeds within a crop affects productivity as a result of the effects of competition.

2. Describe the lifecycles of perennial and annual weeds, and how these are adapted to reduce competition.

3. Explain the damaging effects of invertebrate pests, together with both fungal and viral diseases on growing crops.

4. Describe the mechanisms of cultural, chemical, and biological control methods in reducing the impact of crop pests and diseases.

5. State the advantages and disadvantages of both biological control methods and integrated pest management schemes in the control of crop pests.

6. Explain the costs, benefits and ethics of animal welfare within a named animal production system in the United Kingdom.

7. State the specific animal behaviours which may be observed indicating that an animal is suffering a high degree of stress.

8. Explain the construction and use of an ethogram in the study of animal behaviour.

9. Explain the procedure involved in conducting a preference test with an animal, together with the type of information it provides towards improving animal welfare.

10. Describe what is meant by motivation in animals, with reference to the satisfaction of essential needs in order to thrive.

3.6 Symbiotic and social interactions between organisms

You should already know:

- The term 'species' describes a group of genetically similar organisms which are able to interbreed and produce fertile offspring.
- Intraspecific competition for resources occurs between members of the same species. Interspecific competition occurs between members of different species.
- Competition, parasitism and predation describe specific biotic factors which are interactions between organisms which can occur within an ecological niche.
- Evolution occurs as a result of the processes of mutation and natural selection.

Learning intentions

- State that a symbiotic relationship is co-evolved occurs between members of two different species.
- Explain the benefits and costs to the organisms involved in a parasitic relationship.
- Describe the ways in which parasites may be transmitted to new hosts.
- State that within a mutalistic relationship, both organisms benefit.
- Describe the origin of chloroplasts and mitochondria and the evolution of eukaryotic cells.
- Explain the way in which social hierarchy and cooperative hunting as mechanisms of social behaviour, reduce the effects of competition.
- Explain how the social behaviours of 'altruism' and 'kin selection' can increase chances of survival.
- Describe the complex structure of social insect communities and the role of social insects within an ecosystem.
- Describe the structure of primate social groups and the impact of ecological niche, taxonomic group and distribution of resources, on this.

Symbiosis

The word 'symbiosis' refers to two different species 'living together' in a close relationship, having adapted to one another as a result of co-evolution.

Parasitism and mutualism are two different types of symbiotic relationship.

Parasitism

In a parasitic relationship, one organism benefits in terms of energy or nutrients (parasite) and the other organism is harmed (host). A parasite can live inside or outside the body of the host, feeding on nutrients often taken from the blood. As a result the host becomes weak, and can eventually die. The parasite must then find a new host in order to complete their lifecycle and survive.

Transmission of parasites to new hosts

Most parasites are obligate, meaning that they depend upon a host for survival. If the host dies, they must find a new host in order to continue their lifecycle. Parasites can be transmitted to new hosts by:

1. Direct contact – for example fleas can transfer from one animal's coat to another when they have direct physical contact.

2. Vector – a second organism which can carry the parasite to another host. The 60 different mosquito species belonging to the group *Anopheles* can transmit infectious stages of the malaria parasite *Plasmodium* to humans, as shown in **Figure 3.6.1**.

3. Resistant stages – a parasite may produce a resistant stage as part of its lifecycle. Tapeworms which live in the human intestine produce resistant eggs which leave the body in the faeces. Sewage farms now sell human waste to farmers as a fertiliser to spread on the land. Grass contaminated with tapeworm eggs may then be eaten by cattle or pigs, which are secondary hosts. Tapeworm cysts in the meat may then be eaten by humans, providing a new host for the parasite.

> ### 🔍 Hint
>
> Parasites can be classified as being either **facultative**, meaning they are able to survive outside a host, or **obligate**, meaning they need a host to survive.

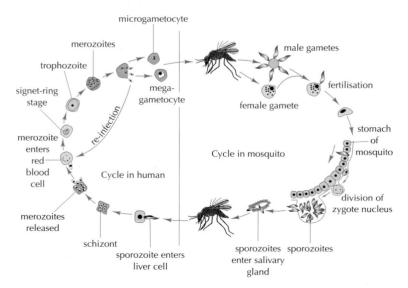

Figure 3.6.1 *Lifecycle of the malarial parasite*

The table below shows some human diseases caused by parasites:

Disease	Parasite	Symptoms and/or transmission
Malaria	*Plasmodium* (4 different species)	Regular fevers. Transmitted to humans by the Anopholes mosquito.
Sleeping sickness	*Trypanosoma* species	Transmitted to humans via insects such as the tsetse fly.
Leishmaniasis	*Leishmania* species	Transmitted to humans via sandflies.
Schistosomiasmis	*Schistosoma* (blood fluke)	Transmitted by flukes (a type of parasitic flatworm whose intermediate host is a water snail).
Tapeworm infection	*Taenia solium* (tapeworm)	Transmitted to humans via infected and undercooked meat. Parasite lives in the human intestine.
Toxoplasmosis	*Toxoplama* species	Weakness and fever. Transmitted to humans by domestic cats. Passed from mother to baby across the placenta.

Make the link

The process of cell respiration is described in Unit 2.

Mutualism

Mutualism describes a relationship between individual organisms of two different species in which both benefit.

For example, lichens are formed as a result of a symbiotic relationship between an algae and a fungus. The algae grows inside the fungus for protection. The photosynthetic algae provides the fungus with glucose for cell respiration, and the fungus provides the algae with water, minerals and protection. Both the lichen and the fungus benefit from this mutualistic relationship.

Additionally, lichens are able to exploit a much wider range of environments than an individual species of fungus or algae.

Hint

Nitrogen fixation by *Rhizobium* bacteria is an important part of the Nitrogen Cycle discussed in National 5.

Figure 3.6.2 *Foliose Lichen species growing on bare rock*

Rhizobium bacteria, which live in the root nodules of leguminous plants such as peas, beans and clover, are another example of mutualism. These nitrogen-fixing bacteria convert nitrogen gas from the air in the soil into ammonia. The nitrogen compound ammonia diffuses out of the root nodules into the soil, and is taken up through the roots of the plant. This provides the plant with nitrogen fertiliser, which increases growth. In return, the plant provides the bacteria in the root nodules with glucose from photosynthesis for bacterial cell respiration, together with a suitable environment in which to live.

Figure 3.6.3 *Root nodules containing Rhizobium bacteria in a leguminous plant*

Endosymbiont theory

The endosymbiont theory proposes that both mitochondria and chloroplasts were once free-living **prokaryote** organisms. The hypothesis is that a prokaryote unicellular organism which was respiring aerobically became engulfed by the plasma membrane of a **eukaryote** cell. Once inside the cytoplasm, the prokaryote cell provided the host cell with energy in the form of ATP, and in return the host cell provided oxygen and glucose. This was the beginning of a mutualistic relationship in which over time, the prokaryote cell evolved into a **mitochondrion**.

Free-living unicellular photosynthetic prokaryotes were then engulfed by the plasma membrane of eukaryote cells containing mitochondria. A mutualistic relationship was then established between the mitochondria and the photosynthetic prokaryote unicellular organisms. The mitochondria provided carbon dioxide and water for photosynthesis, while in return the photosynthetic prokaryotes provided both oxygen and glucose for aerobic respiration. Eventually, the photosynthetic prokaryote cells evolved into **chloroplasts**.

Evidence of the prokaryotic origins of mitochondria and chloroplasts may be found today, as both contain circular chromosomes found in prokaryote cells.

📖 Prokaryote

A living cell which does not contain a membrane around its nucleus.

📖 Eukaryote

A living cell which contains a membrane around its nucleus.

📖 Mitochondria

Structure within the cytoplasm of eukaryotic cells where aerobic respiration takes place, generating ATP.

📖 Chloroplast

Structure within the cytoplasm of plant cells where photosynthesis takes place.

⁙ Make the link

The endosymbiont theory explains the evolution of the mutualistic relationship of chloroplasts and mitochondria within a living plant cell. It also illustrates the way in which the processes of cell respiration and photosynthesis are interdependent within a plant cell. The process of respiration is discussed in Unit 2, and the process of photosynthesis is discussed earlier in this unit. It is important to remember how these two processes are linked together in a plant cell, rather than thinking of them as separate cell processes.

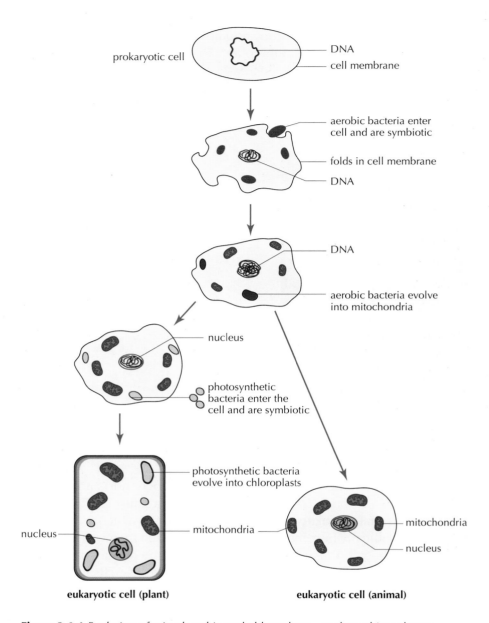

Figure 3.6.4 *Evolution of mitochondria and chloroplasts – endosymbiont theory*

Social behaviour

Animals that live as part of a social group have evolved patterns of behaviour which increase their chances of survival, through reducing the effects of both predation and competition.

These patterns of behaviour include cooperative hunting, social or dominance hierarchies, co-operative defence against predation, altruism and kin selection.

Cooperative hunting

Some species organise themselves into groups in order to find and kill prey. This reduces competition between individuals and increases the chances of obtaining larger prey by working together. At the end of the hunt, the kill is shared between members of the group. This means that each individual benefits from working as part of a team.

For example, chimpanzees are omnivores that eat meat together with leaves and fruit. They sometimes hunt small monkeys, such as the Colobus monkey, for meat. It would be very difficult for one chimpanzee to catch a small monkey in the trees of the forest, however, working as part of a team greatly increases the chances of a kill.

Male chimpanzees form the hunting party, with each individual taking on a specific role. Once a monkey has been identified as the prey, the 'drivers' climb trees and chase the monkey ahead of them. 'Blockers' climb trees ahead of the monkey to prevent it moving forward and 'chasers' keep the monkey moving forward. Climbing trees ahead of the monkey, 'ambushers' drive the monkey back towards the 'chasers' closing in on the prey. As the circle of chimpanzees tightens, all individuals move in quickly for the kill.

The meat is shared between all members of the hunting group, according to their social position. The remaining meat is taken down to the ground for females and young, who observe the hunt from a distance.

Figure 3.6.5 *Colobus monkey, one source of prey for chimpanzees*

Figure 3.6.6 *Ambusher waits in trees for monkey to approach*

Another example of cooperative hunting may be seen in bottlenose dolphins. One dolphin (called a 'driver') drives a small shoal of fish towards a line of five or six dolphins, who are waiting ahead. The fish are blocked by the line of dolphins, unable to move forward or back. The result is that they move upwards toward the surface of the water. The 'driver' dolphin slaps the water with its tail fluke, making a great deal of noise and vibration, which travels through the water. The frightened fish then leap upward, out of the water and into the open mouths of the waiting dolphins. Each member of the hunting group benefits from working cooperatively, in terms of food.

Figure 3.6.7 *Bottlenose dolphins working together to block the path of an oncoming shoal of fish*

Social hierarchy

The position of an individual within a social hierarchy, or dominance hierarchy, indicates the priority each individual takes over another in terms of access to food, shelter and mates, for example.

Each individual has a specific place within a linear hierarchical ranking structure, which is allocated through the outcome of aggressive behaviours between individuals.

The individual at the top of a social hierarchy has dominance over all other individuals, and has first choice of food and mates. All other individuals are dominant to those below it in the hierarchy, and subordinate to those individuals who are above.

A social hierarchy may be seen in animals such as wolves, chimpanzees, horses, and chickens.

Social hierarchy in chickens

Position 1 – Chicken dominant to all others; first choice of food and mates.

Position 2 – Chicken submissive to chicken in position 1; second choice of food and mates.

Position 3 – Chicken submissive to chickens in positions 1 and 2; third choice of food.

Position 4 – Chicken submissive to chickens in positions 1, 2, and 3; has fourth choice of food.

Position 5 – Chicken submissive to chickens in positions 1, 2, 3, and 4; has fifth choice of food.

Position 6 – Chicken at end of hierarchy, submissive to all others in group; only has access to food left by chickens in positions 1 to 5.

Positions within the hierarchy are determined by aggressive pecking behaviour, which results in one bird becoming dominant over the other.

Advantages of social hierarchy

The advantages of a hierarchical structure within a social group are:

- Reduced fighting, resulting in a lower chance of injury to individuals
- Energy being conserved
- Individuals at the top of the hierarchy are more likely to pass on favourable genes to the next generation

Co-operative defence against predation

Flocking behaviour

Large groups of animals have better chances of defending themselves against attack by a predator than a small group, or individual animals outwith a group.

The larger the social group, the less chance each individual member of the group has of being attacked by a predator. This is known as the dilution effect and can be seen in large flocks of starlings going to roost. The larger the flock, the less likely an individual is to be predated by falcons that await the flock.

Figure 3.6.8 *A starling murmuration – an example of flocking behaviour*

Vigilance behaviour in barnacle geese

Within a flock of several hundred geese grazing in a field, most will have their head lowered feeding. Some geese however remain vigilant, with their heads raised scanning the horizon for possible signs of a predator while the others feed. If a potential predator is seen, the vigilant geese raise the alarm, and the whole flock takes to the air for safety.

Figure 3.6.9 *One barnacle goose displaying vigilance behaviour while the others graze*

'Mobbing' of predators

Figure 3.6.10 *Arctic terns displaying mobbing behaviour*

'Mobbing' of a predator occurs when it is surrounded by a large number of adult prey animals, in order to protect their young.

For example, Arctic Terns nest in colonies on some Scottish beaches on ground above the high water mark. People and dogs walking on the beach are seen as a potential danger to the newly hatched chicks amongst loose rocks on the sand. Adult birds fly together to 'mob' the potential predators by dive bombing and violently pecking the heads of intruders, while emitting high pitched warning screeches. If walking on beaches where Arctic Terns nest, it is advisable to wear a hat, or better, follow the advice and keep away from tern colonies – the birds are specially protected.

Sheep will leave their lambs and surround a fox, head butting the predator and making a great deal of noise in order to drive it away from their lambs.

Social defence of a mobile group

Figure 3.6.11 *Females and young travel towards the inside of the social group for safety*

Animals travelling in groups, seeking out new environments in which to live are susceptible to attack by predators.

Troops of baboons travel with the dominant males, females and young towards the inside of the group for safety. Individuals holding a low position within the social hierarchy travel on the perimeter of the group, and raise the alarm if a predator is approaching, as shown in **Figure 3.6.11**.

Altruism and kin selection

Altruism is the term used to describe unselfish behaviour. An individual (donor) may engage in a behaviour which is of advantage to another (recipient) within a social group. The behaviour benefits the recipient, but results in a disadvantage to the donor. The chances of survival are increased for the recipient, but decreased for the donor. However, this debt *may* be repaid by the recipient to the donor at another time. This is called reciprocal altruism. For example, a vampire bat at a roost site will share a blood meal with another hungry bat, who will return the favour on another occasion, by sharing its blood meal with the original donor bat.

When food is in short supply chimpanzee mothers will give their food to their young, meaning that the mothers go hungry. This is an altruistic behaviour.

Altruism may have evolved from kin selection, a form of natural selection where an individual protects the survival of their own genes in future generations by giving priority to the protection of family members first above all others.

Individuals may sacrifice their own lives in order to save close relatives, preserving the genes they both share for future generations.

Female chimpanzees that cannot reproduce will take an active role in the care of the offspring of a close female relative.

Social insects

Both altruism and kin selection have contributed to the evolution of social structure within insect societies. Complex social structures can be found in colonies of bees, ants, wasps and termites.

Social structure of a honey bee colony

A typical bee colony consists of around 50,000 bees. However, only the **queen bee** lays eggs, while the female offspring of the queen (called **worker bees**) keep the hive clean and defend it against predators, rather than breeding themselves. This is an example of both altruistic behaviour and kin selection. Worker bees take care of the queen bee as they are all genetically related to her, and they make her wellbeing a priority because genes they share with her are more likely to be passed on to offspring.

A **drone bee** is a male which fertilises the eggs laid by the queen, and works to ensure the survival of the larvae. This is an example of kin selection, as the larvae are the offspring of the drone bees.

Drones provide food for the larvae, clean breeding cells within the hive, build new breeding cells, and help to defend the hive against predators.

> ### 📖 Word bank
>
> **Queen bee**
> Only female who lays eggs in a colony.
>
> **Worker bee**
> Sterile female bees.
>
> **Drone bee**
> Males which fertilise the eggs of the queen bee.

Figure 3.6.12 *Drone bees cleaning breeding cells of hive*

 Hint

Find out about social structure in other insect species, such as termites or ants.

Make the link

Biological yield and economic yield are discussed in Unit 3.3

Keystone species

Social insects support the intricate balance of an ecosystem, and are often regarded as keystone species. This reflects the importance of their contribution.

Decomposition is a key part of the nitrogen cycle. The species *Porcellio scaber* or common rough woodlouse, are decomposer organisms, eating and digesting organic matter which can then be further broken down by bacteria to release nitrogen compounds into the soil. Common rough woodlice have a key role in the process of decomposition within their ecosystem by maintaining the nitrogen cycle. They are a keystone species.

The garden bumblebee (*Bombus hortorum*) is also a keystone species within its ecosystem, transferring the male gamete pollen to the female stigma of flowers, and thereby ensuring the process of fertilisation as part of sexual reproduction in plants. This results in the production of seeds. Pollination by bees ensures genetic diversity in plants through outbreeding. Outbreeding is explained fully in Unit 3.4.

Ecosystem services

The activity of social insects within an ecosystem can result in processes or products which are of economic benefit to humans.

The pollination of yellow oilseed rape flowers produces seeds, which are then crushed to produce rapeseed oil for food manufacture. The production of economic yield is dependent upon the crop being pollinated by bees.

African mango farmers are able to increase their crop by 75% using weaver ants. These ants feed on the fruitflies which attack the mango fruit, making it unsaleable.

This is an example of biological pest control using a social insect.

Behaviour of primates

Primates are placental mammals which include lemurs, chimpanzees, gorillas and humans. They have two mammary glands, five-digit limbs, an opposable thumb, claws modified into flat nails, binocular vision and a large brain relative to body size.

Figure 3.6.13 *Mother and baby Gorilla*

Parental care

Primates usually give birth to single offspring which are helpless, and require a high level of parental care for a number of years as they grow. Parental care involves feeding the infant, keeping it warm and protecting it from danger. Primates carry infants when moving from place to place.

While being cared for by parents, young primates have the opportunity to learn social behaviours essential for future survival through play and verbal communication, modifying their behaviours appropriately, as guided by the parents.

Reinforcing social structure

In primate groups a social hierarchy exists, which is established by display and threat behaviour. The framework and advantages of a social hierarchy are discussed in the previous section. Each individual within the social hierarchy is dominant over those below, and submissive to those above. This reduces conflict and conserves energy.

Threat displays occur when two individuals compete for the same resources, and attempt to intimidate one another. For example, in chimpanzees threat displays include walking upright on two legs, slapping hands together, body hair erect, teeth exposed, and dragging a large branch. Eventually, one individual will back down, avoiding unnecessary conflict, and display **appeasement behaviours** such as crouching, holding out hand, and proceeding to groom the coats of others in the group.

 Hint

Antagonistic behaviours, including fighting and running away, are seen during situations of conflict.

📖 Appeasement behaviour

Behaviour that reduces aggression in another member in the same species.

Formation of alliances

In primates, strong social bonds, or alliances, form between some individuals. These alliances are important in maintaining the position of an individual within a social hierarchy, or providing the opportunity for an individual to move up to a higher position on the social ladder. These alliances are strengthened and maintained by grooming sessions, where one individual removes the parasites from the coat of another.

Influence of external factors on social structure

The complex social structure of groups of primates differs according to the influence of external factors, such as ecological niche, resource distribution and taxonomic group.

Ecological niche

An ecological niche may be thought of as the role a species has in the environment, together with how it might interact with and use resources.

Primate species are adapted to living in trees and inhabit a wide range of ecological niches, the structure of which can influence the social organisation of a specific group.

For example, the Black Spider monkey (*Ateles paniscus*) forages over large areas to find fruit to eat. When fruit is in short supply, spider monkeys exhibit a 'fission–fusion' social system, meaning they can leave the main social group in order to forage, rejoining the group again later.

In addition to the core community of spider monkeys, smaller subgroups form which travel together within the range of the core community, to find food. These small foraging groups are temporary, changing their composition frequently throughout the day.

Resource distribution

A social group of primates will forage over a specific area each day, known as their home range. The size of this area can vary depending upon the preferred food type of the primate species. If a specific food type is required such as fruit berries for example, then the foraging range will be large as the density of fruit berries is small. If a primate species such as the chimpanzee is able to utilise a wide range of vegetation, then the foraging area will be small.

Most primates are omnivores, and require some meat in addition to plant material. The size of the foraging area will depend on the presence of small prey animals for meat.

Taxonomic group

Primates belonging to the same taxonomic group are found to share both a similar social structure and an ecological niche. This is not the case for primates from very different taxonomic groups, which exploit different ecological niches.

For example, lemurs belonging to the taxonomic group *Lemuroidea* live in large social groups on the African island of Madagascar. They use their hands and feet to move through trees as they cannot grip branches with their tails. Lemurs spend most of their time on the ground, foraging for fruit and eating leaves, treebark and sap.

The black and white Colobus monkey lives in small groups in equatorial Africa and belongs to the taxonomic group *Simiformes*. The Colobus monkey has a very long tail for gripping branches, and lives in the tree tops. It will only descend to the ground if the path of travel in the tree canopy is blocked, and feeds on vegetation from the tree canopy.

As the lemur and Colobus monkey belong to two different taxonomic groups, they possess different social structures and exploit different ecological niches. Lemurs live in large groups with a distinct social hierarchy, but Colobus monkeys live in small groups without a definite hierarchical structure.

Figure 3.6.14 *Ring-tailed lemur*

Hint

Understanding the social structures that exist within groups of animals can improve the welfare of animals within farming systems.

For example, introducing a new group of cows bought from another farm into an existing herd would immediately result in aggressive displays of behaviour, in order to establish the position of the newcomers within the social hierarchy of the herd. If the introduction of new animals into a herd takes place within a confined space, such as a cattle shed, the chance of injury during aggressive displays of behaviour is high, from obstacles such as gates, troughs and metal bars.

However, knowledge of the social hierarchy within a cattle herd means that it is now standard practice for farmers to introduce new animals to a herd in an open, spacious field which has been previously checked for potential hazards such as loose fence wire, and empty overturned feeding troughs.

Cattle are able to engage in challenges for a position within the social hierarchy, and have the space to move away from others when displaying submissive behaviours, without risk of injury.

GO! Activity 3.6.1 Work individually to:

Restricted response

1. **a)** Explain what is meant by a symbiotic relationship between two living organisms. 1

   b) Describe the way in which a parasitic relationship differs from a mutualistic relationship. 1

2. List the advantages of cooperative hunting within a social group of animals. 3

3. Explain the ways in which kin selection can increase the survival chances of individuals within a species. 2

4. **a)** Barnacle geese, which overwinter in Britain, are preyed upon by foxes while grazing in fields on farmland.

   A group of geese was observed across a large area of farmland, and the **average values** of data gathered are presented in the following table:

Observations	Male geese	Female geese
% in whole observed group	20 %	80 %
% killed by foxes	68 %	32 %
% found positioned on perimeter of flock	75 %	25 %
Distance to nearest neighbour (meters)	8·1 (m)	3·8 (m)
% time spent with head up scanning horizon	7·8 %	10·4 %

   From the data in the table:

   i. Calculate the number of female geese killed by a predator within a flock of 523 birds; 1

   ii. Give **two** possible reasons why male geese are more likely to be killed by a predator than female geese; 2

   iii. Display the information given in the table as a bar chart. 3

Extended response

1. Describe the social structure that exists within a colony of bees. 6

2. Discuss social behaviour in primates. 9

GO! Activity 3.6.2 Work in pairs to:

Jane Goodall is a world-renowned authority on chimpanzee behaviour. Locate the website of the Jane Goodall Research Institute at www.janegoodall.org. Click on 'Chimpanzees' on the tool bar. Select information from the sections on using tools, biology and habitat, and communication. Use this information to **make a 5 slide Powerpoint presentation or wall display**.

Alternatively, **prepare an illustrated talk** for the class on a parasite which uses humans as a host. Research **one** human parasite from the following list:

- Plasmodium faciparum – malaria
- Toxoplasma gondii – toxoplasmosis
- Tapeworm – Diphyllobothrium
- Fluke (flatworm) – Fusciola hepatica
- Enterobius (pinworm)

Your talk should include information on:

- The lifecycle of the parasite
- Classification of parasite – obligate or facultative
- Methods of parasite transmission between hosts
- Effect on host organism – symptoms
- Control and treatment of parasite infestation

GO! Activity 3.6.3 Work in groups to:

Explore the area around your school, looking for lichens on walls, trees, shrubs and possibly on any concrete walls. Examine each species of lichen closely with a magnifying lens. Record the different types of lichen found by sketching or taking a digital image. Make a note of the colour of the lichen species, and the shape of the thallus of 'leaves' of the organism. Record beside each species an 'abundance rating' from the scale.

Lichen species covers most of area = 1

Lichen species covers about 75% of area = 2

Lichen species covers 50% of area = 3

Lichen species covers very little of area = 4

Identify each lichen species recorded using the website www.britishlichens.co.uk. Compare your findings with those of other groups, and construct a wall chart displaying the class results.

After working on this chapter I can:

1. Describe the difference between a symbiotic and parasitic relationship, in terms of benefits and costs to the organisms involved.

2. State an example of both a symbiotic and a parasitic relationship.

3. Describe the lifecycle of the malaria parasite, and the role of the mosquito as a transmission vector.

4. Explain the endosymbiont theory, which describes the evolution of the mutualistic relationship between mitochondria, chloroplasts and eukaryotic cells.

5. Describe the way in which the position within a social hierarchy is established, and the advantages of a hierarchical structure within a social group.

6. Describe, using a named example, the benefits of cooperative hunting within a social group.

7. Explain how animals have evolved behaviours to defend themselves against predation.

8. Define the terms 'altruism' and 'kin selection', explaining how each behaviour may increase the survival chances of a species.

9. Describe the social organisation within a colony of honeybees, and the role of the honeybee as a 'keystone' species.

10. Describe the social structure which exists within a group of primates, and how individual roles within the social structure are determined and maintained.

11. Explain how the role of an individual primate within an ecosystem, distribution of food as a resource, and species of primate can impact on the social structure of a particular group of primates.

3.7 Biodiversity

You should already know:

- Biodiversity is the genetic variation which exists within species or the variety and abundance of species themselves.
- Predation and grazing are biotic factors which can affect **biodiversity.**
- Abiotic factors such as temperature and pH also affect biodiversity.
- Human activities are often threats to biodiversity.

Learning intentions

- Describe a 'mass extinction' event and the subsequent recovery of biodiversity.
- State how an estimate of 'extinction rate' may be obtained.
- Describe how biodiversity can be measured by focusing on genetic diversity, species diversity and ecosystem diversity.
- Explain the impact of exploitation and recovery of populations on genetic diversity.
- Describe habitat fragmentation and how its impact on biodiversity can be reduced using habitat corridors.
- Explain the impact of introduced, naturalised, and invasive species on biodiversity.
- Explain the impact of climate change on biodiversity.

Mass extinction

Mass extinction events

The extinction of a large number of species within a relatively short period of geological time is known as a mass extinction event. These events may be due to environmental change, which occurs too quickly for species to evolve adaptations in order to cope. A meteor striking the Earth, causing immediate changes to the atmosphere is an example of this.

A mass extinction event results in a catastrophic loss of biodiversity.

Using information from fossil records, five mass extinction events have been identified within the Earth's history. The largest of these was the Permian extinction, which occurred 245 million years ago. As a result, 90% of all marine species were lost, together with about 75% of all terrestrial species. Volcanic eruptions into the atmosphere might have been the cause as clouds of dust blocked out the sunlight, reducing photosynthesis and causing the collapse of food chains.

Second mass extinction event

Another second mass extinction event occurred in the Cretaceous period about 65 million years ago. This event resulted in the well-known loss of terrestrial dinosaurs, together with around half of all marine species of animals.

Holocene extinction event

This mass extinction event is speculated to be underway today, and began with the evolution of humans on Earth. Human activities such as hunting, building roads and cities, agriculture and industrial pollution are endangering many species of large animal or megafauna due to the degradation of the ecosystems in which they live. Loss of habitat and competition with humans for natural resources are two examples of the ways in which megafauna have suffered in recent times. Elephants, rhinoceros, tigers and mountain gorillas are just some of the megafauna in danger of extinction today.

Figure 3.7.1 *Images of human pollution effects on the environment*

Extinction rate

The total number of species that become extinct within a specific geographical area over a specific period of time, is called the extinction rate.

Calculating a value for an extinction rate is easier with large animals and plants which can be seen and recorded easily. It is a difficult calculation to make for insects, for example, which are difficult to observe and count accurately. In both cases, the calculation of an extinction rate can only be an estimate, at best.

Recovery of biodiversity

Once the surviving species of an extinction event have begun to recover, they undergo the processes of natural selection and speciation, which take place over several millions of years. During this time new species will begin to emerge, which leads to a gradual increase in biodiversity over time. These new species are able to radiate out and exploit vacant niches left behind following an extinction event.

Importance of biodiversity

Conserving biodiversity is important for maintaining a wide variety of genetically diverse species on earth, and preventing the permanent loss of valuable and unique genes. Maintaining the biodiversity of the planet is an important legacy for future generations of all species.

>
> **Hint**
> Biodiversity is important for sources of food, economic value, and the stability of ecosystems.

Measuring biodiversity in an ecosystem

Biodiversity is difficult to measure. There are three aspects of biodiversity which may be measured within an ecosystem:

1. **Species diversity** – can be estimated by recording the number of different species of living organisms within an ecosystem, together with their relative abundance. The greatest species diversity is usually found within insects.

2. **Genetic diversity** – this measure involves examining a specific population of animals or plants, and recording the number and frequency of alleles within the population gene pool.

3. **Ecosystem diversity** – a measure of the number of different ecosystems present within a specific geographical area.

> **Hint**
> Remember, sampling techniques used to estimate species diversity include:
> * Quadrants
> * Pitfall traps
> * Tree beating
> * Line transects
> * Sweep netting

Threats to biodiversity

Formation of habitat islands

Natural habitats containing a variety of animal and plant species can be broken up into fragments or islands, by either geological or human activity. This results in some species becoming isolated within habitat islands, surrounded by an ecosystem which is unfamiliar to the trapped species.

Species diversity is then reduced due to the increased isolation of the habitat, and the decrease in size of the total habitat area.

Process of habitat fragmentation

The fragmentation - or break-up - of a habitat into islands involves the following steps:

- Isolation of discrete fragments
- Break-up of one area of habitat into several smaller individual areas.
- Degradation of the edges of a fragment, causing further decrease in the size of the habitat fragment, leading to a decrease in the interior-to-edge ratio.

Some species have adapted to living on the edge of habitat fragments, and these are called edge species. They may, however, invade the interior of the habitat fragment and compete with interior species for resources, such as food and space. For instance, rabbits will colonise the outer edges of a habitat fragment where there is plenty of grass to eat. They may then move into the interior and compete with other herbivores for available food.

Habitat corridors

A natural or man-made area of habitat may connect two habitat fragments, forming a bridge (or corridor), which allows the movement of individuals between populations and helps restore biodiversity.

For example, the construction of new motorways in Britain include concrete tunnels in the foundations, which allow mammals such as hedgehogs and badgers to move safely between habitat islands on either side of the road, and avoid being run over as they try to cross it.

Habitat corridors can make an important contribution to the stabilisation of populations through colonisation, migration and interbreeding.

- Colonisation – involves the movement of animals through the habitat corridor, allowing the exploitation of new resources. Habitats may be re-colonised that have previously suffered an extinction event.
- Migration – a habitat corridor provides an undisturbed migration route, allowing the free movement of migratory species.
- Interbreeding – movement of animals through the corridor provides greater opportunities for outbreeding, resulting in an increase in genetic diversity.

Figure 3.7.2 *Habitat corridors allow animals such as hedgehogs to cross under the road safely*

Figure 3.7.3 *Animal bridges help to reduce habitat fragmentation*

Human exploitation

Economically valuable food species, such as North Sea cod have been overexploited by human fishing activities, resulting in the population size of cod becoming low, raising the possible threat of extinction. Small, young fish are caught, killed and thrown back into the sea as they have little value. These 'discards' die before reaching their reproductive potential.

The **over-explotation** of fish stocks has been recognised and attempts are being made to allow populations to recover through the introduction of fish quotas for fishing vessels, and a limit to the time boats can spend fishing at sea.

Bottleneck effect

A population may be drastically reduced as a result of an event, shrinking the available gene pool. For example, a flooded woodland may wipe out most of a population of badgers living in setts underground. The few badgers that survive the flood will now share a very small gene pool, as many alleles will have vanished with those who did not survive.

This means that it will be more difficult for the badgers to adapt to changes in climate, as the genetic variation which underpins evolution has been lost due to the flooding of the woodland ecosystem.

Introduced, naturalised and invasive species

The movement of a **species** from one location to another by humans can have a devastating effect on the fine balance of food webs within an ecosystem.

The New Zealand flatworm (*Arthurdendyus triangulates*) is a **species** which was introduced to Britain through contamination of the compost of pot plants imported from New Zealand in the 1960s. When the plants died, the compost was discarded into gardens, releasing the flatworm into a new environment.

The flatworm quickly adapted to the new soil conditions, feeding mainly on earthworms, and became **naturalised**. It competed fiercely with other animals which depended upon earthworms for food, such as hedgehogs and blackbirds. Many gardens which are infected with the new Zealand flatworm no longer have hedgehogs for this reason. Population numbers of earthworms, and subsequently hedgehogs, in Britain are beginning to fall as a result.

📖 Over-exploitation

This occurs when resource stock is placed under too much pressure, leaving too few mature individuals to produce the next generation.

📖 Species

Individuals who are genetically similar, and are able to interbreed to produce fertile offspring.

📖 Naturalised

Describes a species that has been introduced into a new environment and is able to establish a wild, self-sustaining population.

🔍 Hint

The New Zealand flatworm is an invasive species, which has reduced the numbers of indigenous species such as the earthworm. This reduces biodiversity.

The New Zealand flatworm has now become an **invasive species**.

In 1992, the New Zealand flatworm was added to Schedule 9 of the Wildlife and Countryside Act, making it an offence to release the flatworm into the wild again if captured.

📖 **Invasive species**

A non-native species that colonises local ecosystems.

Climate change

The greenhouse effect is caused by heat radiation that is unable to escape from the surface of the Earth due to a thick atmospheric blanket. This blanket of gases, including carbon dioxide (CO_2) and methane (CH_4), can cause a rise in global temperature and a major change in weather patterns. Evidence for a rise in global temperature can be found in the disappearance of Arctic sea ice, as shown in the following photographs.

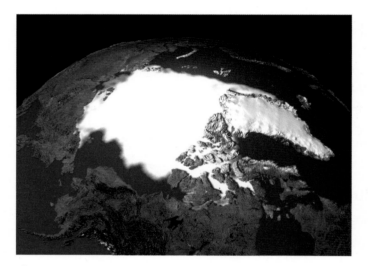

Figure 3.7.4 *Arctic ice minimum extent in 1979*

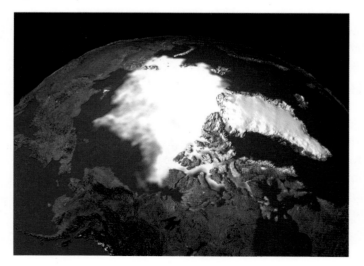

Figure 3.7.5 *Arctic ice minimum extent in 2003*

Many species of living organisms are sensitive to ambient temperature, and if it changes significantly within an ecosystem, a different species will become dominant. A change in average yearly rainfall may result in the extinction of some species, while other species may thrive.

Some progress has already been made toward reducing the effects of climate change by burning less fossil fuel. Increasingly, we are looking instead at alternative forms of energy such as wind, solar and wave. Many countries have now committed to reducing their carbon emissions, providing hope for the future survival of biodiversity.

⚛ Make the link

A mass extinction event could be caused by a volcanic eruption filling the global atmosphere with dust. This would block out the majority of available sunlight and have an impact on the absorption and action spectrum of photosynthesis, as discussed in Unit 3.2. Producers in food chains would become less productive, reducing energy flow through ecosystems. Plant productivity is discussed in Unit 3.3.

GO! Activity 3.7.1 Work individually to:

Restricted response

1. **a)** Give one possible cause of a mass extinction event, and how it would affect biodiversity. 2

   b) Give reasons why megafauna are in danger of becoming extinct. 3

   c) Explain what is meant by the term 'extinction rate', and state how it might be calculated. 2

   d) State the three components of biodiversity which collectively provide a measure of total biodiversity within an ecosystem. 3

2. **a)** The following is an extract from a marine biology research report.

   Read the extract carefully and answer the questions which follow:

   Fishing in the North Sea is critically affecting stocks of native cod. Presently, there are approximately 40–60,000 tonnes of breeding cod in the North Sea. This number may also be as a result of fishing for other species such as haddock and plaice, where cod may be unintentionally caught alongside other species.

   For cod stocks to become stabilised, it is estimated that there must be a minimum of 80,000tonnes present in the North Sea, and 160,000 tonnes to ensure sustainability. This number of of cod was last achieved twenty years ago.

(continued)

At around four years of age, cod become mature enough to reproduce. However, fishing boats are catching nearly all three year old fish, and about half of all two year olds. Marine biologists fear that in the near future, cod stocks may disappear completely from the North Sea.

   i. State the number of years since stocks of North Sea cod exceeded 80,000 tonnes. 1

   ii. Give possible reasons for the rapid decline in fish stocks. 1

   iii. Calculate how many more thousand tonnes of spawning cod in the North Sea are needed to reach the minimum of 80,000 tonnes, in order to stabilise the cod population. 1

   iv. North Sea cod are able to mature and reproduce when they are four years old. Describe the effect of removing all three-year-old fish and half of the two-year-old fish from the population. 1

b) The following image shows the extent of widely established invasive non-native species in freshwater, marine and terrestrial environments, from 1960 to 2008:

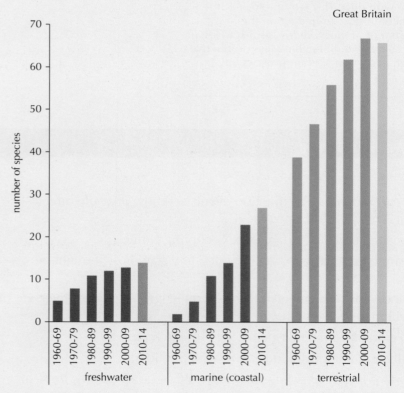

Figure 3.7.6 Source *Centre for Ecology & Hydrology, British Trust for Ornithology, Marine Biological Association and the National Biodiversity Network Gateway. (UK DEFRA)*

   i. Identify the type of ecosystem which has more than 50% of the total area hosting invasive species each year, from 1960 to 2008. 1

   ii. Identify the year in which the number of invasive species established on 10–50% of the freshwater system area was double that of the marine ecosystem between, 1960 and 1980. 1

iii. Calculate the percentage increase in the number of invasive species established on more than 50% of land area within the terrestrial ecosystem from 1960 to 2008.　　　　1

iv. Areas greater than 50% inhabited by invasive species are much lower for freshwater and marine ecosystems, compared to terrestrial ecosystems. Give possible reasons for these results.　　　　1

Extended response

Describe the formation of habitat islands, and explain how the subsequent reduction in biodiversity can be overcome by the formation of habitat corridors.　　　　8

Activity 3.7.3 Working in groups:

The following table lists a number of invasive species of animal present in Britain today. Choose one invasive species to study, and use the research information to **produce a short play** explaining the effects of the invasive species on biodiversity.

The play should include:

- How the species was transported by humans to a new ecosystem;
- the ability of the species to adjust to the new environment;
- the effect of competition on the native species within the ecosystem;
- the effect of local biodiversity of the invasive species;
- how the effects might be controlled or reduced.

Arthurdendyus triangulatus (New Zealand flatworm)	*Harmonia axyridis* (Harlequin ladybird)
Branta canadensis (Canada goose)	*Oxyura jamaicensis* (Ruddy duck)
Mustela vison (American mink)	*Muntiacus reevesi* (Reeves' muntjac)
Sciurus carolinensis (Grey squirrel)	*Rattus norvegicus* (Brown rat)

After working on this chapter I can:

1. Explain the impact of a mass extinction event on biodiversity.

2. Understand the possible reasons for a mass extinction event to occur, including human activity.

3. State how to estimate the extinction rate of a species.

4. Describe how biodiversity can be measured by examining genetic diversity, species diversity, and ecosystem diversity.

5. Understand the consequences of human over-exploitation of food species on genetic diversity, and the importance of species recovery.

6. Describe the process of habitat fragmentation, and the impact on biodiversity, in terms of degree of isolation and area of habitat.

7. Explain how habitat fragments may be linked by habitat corridors acting as 'bridges' between isolated habitats.

8. State the advantages of habitat corridors to the conservation of genetic diversity.

9. Describe the difference between 'introduced', 'naturalised' and 'invasive' in terms of exotic species, which are transported into a new ecosystem by the action of humans.

10. Explain the impact of increased global temperature and rainfall on biodiversity of plant and animal species.

Answers: individual activities

Unit 1

Chapter 1.1

1. **a)** Nucleotide base, sugar and phosphate (all 3 = 2, 2/1 = 1).

   b) Weak hydrogen bonds (1) between complementary bases (1).

   c) Double helix

   d) i. Pairs of nucleotides which are able to join together with hydrogen bonds. For example, A only pairs with T, and C only pairs G

   ii. Parallel strands which run in opposite directions **or** one strand runs 3′ to 5′ and the other runs 5′ to 3′.

2. **a)** The concentrations of adenine and thymine are similar in both organs, as are the concentrations of cytosine and guanine.

   b) The sperm had 0·29 adenine and 0·30 thymine. The liver had 0·19 guanine and 0·18 cytosine. The concentrations of each base pair should be the same.

3. Label axes and scales (1). Plot data and straight lines joining data (1).

Extended response (any 6)

- made up of nucleotides
- nucleotide has a deoxyribose sugar, phosphate and a base
- strands linked by sugar-phosphate backbones
- different bases are adenine, guanine, cytosine and thymine
- bases pair A-T or C-G
- bases are complementary
- antiparallel strands with sugar at 3′ and 5′
- double helix

Chapter 1.2

1. linear chromosomes

2. **a)** Plant, animal, fungi yeast (1).

   b) Bacteria, Archaea (1)

3. Mitochondria, chloroplast

4. Tightly coiled (1)

   packaged with associated proteins (1).

Extended response (any 6)

- Circular chromosomal DNA in prokaryotic cells
- Plasmids in prokaryotic cells.
- Circular plasmids in yeast.
- Circular chromosome in mitochondria of eukaryotic cells
- Circular chromosome in chloroplasts of eukaryotic cells.
- DNA in the linear chromosomes of the nucleus of eukaryotic cells
- DNA is tightly coiled in eukaryotic cells
- DNA packaged with associated proteins in eukaryotic cells.

Chapter 1.3

1. ATP, enzymes, DNA template, free DNA nucleotide bases, primers. (Any 4 = 2 / Any 3 or 2 = 1)

2. **a)** Small section of DNA required for replication to begin.

   b) Enzyme which adds complementary nucleotides to the deoxyribose (3′) end of a DNA strand.

   c) Fragments of DNA are joined together by ligase.

3. Leading strand replicated continuously (1)

   Lagging strand replicated in fragments. Ligase is used to join the fragments (1).

4. To amplify regions of DNA.

5. Any of the options from below.

   - DNA sequencing
   - Genetic mapping studies (for example, the Human Genome Project)
   - Forensic and parentage testing
   - Sex determination in pre-natal cells
   - Classification of species into taxonomic groups based on ribosomal RNA genes sequences
   - Screening for and diagnosis of genetic disorders (such as cystic fibrosis)

Extended response (any 5): DNA replication

- DNA unwinds into two strands
- Primer needed to start replication

- DNA polymerase adds DNA nucleotides to 3′ (deoxyribose) end of strand
- DNA polymerase adds DNA nucleotides in one direction
- Leading strand replicated continuously, the lagging strand in fragments
- Fragments joined by ligase

Extended response (any 4): PCR

- DNA heated to 95°C to separate strands
- Cooled to 50°C to allow primers to bind to target sequences
- Primers are complementary to specific target sequences at the two ends of the region to be amplified.
- Heat-tolerant DNA polymerase then replicates the primed region of DNA at 70°C
- Repeated cycles of heating and cooling amplify this region of DNA

Chapter 1.4

1. A process involving transcription and translation where DNA sequences are used to direct production of proteins.
2. RNA polymerase.
3. Nucleus.
4. Exons are coding (1); introns are non-coding (1).
5. The introns of the primary transcript of mRNA are removed in RNA splicing (1).

   The exons are joined together to form mature mRNA (1).
6. Ribosome.
7. Anti-codon.
8. Peptide.
9. Cutting and combining polypeptide chains (1)

   Adding phosphate or carbohydrate groups to the protein (1).

Extended response

10. - different amino acid sequences
    - different locations for H bonds to link chains
    - other interactions between amino acids give different folding
      (1 each = 3)

11. • RNA single strand and DNA double strand

• RNA has ribose or DNA has deoxyribose

• RNA has uracil in place of thymine formed in DNA (1 each = 3)

12. mRNA (messenger) (1) carries a copy of the DNA code from the nucleus to the ribosome (1). It has a linear form and groups of three bases known as a codons (1).

(Any 2)

rRNA (ribosomal) (1), (along with ribosomal protein), forms (protein-synthesising organelles called the) ribosomes (1).

tRNA (transfer) (1) molecules each carry a specific amino acid (1) (and are involved in the second part of protein synthesis). They have a folded shape and an anti-codon, a group of three bases (1), each attaches to a different amino acid at an attachment site (1).

(Any 2)

13. a) Transcription (any 5)

• DNA unzips/hydrogen bonds break/ DNA strands separate.

• RNA polymerase adds to complementary DNA bases.

• Guanine pairs with cytosine, uracil pairs with adenine. (Not base letters.)

• RNA polymerase can only add nucleotides to the 3′ end of growing mRNA molecule.

• Primary transcript formed.

• Coding regions are exons, non-coding regions are introns.

• Splicing – introns cut out, exons spliced together to form mature transcript.

b) Translation (any 5)

• Translation of mRNA into a polypeptide by tRNA at the ribosome.

• tRNA folds due to base pairing to form a triplet anticodon site and an attachment site for a specific amino acid.

• Triplet codons on mRNA and anticodons translate the genetic code into a sequence of amino acids.

• Start and stop codons exist.

• Codon recognition of incoming tRNA / anti codon.

- Peptide bond formation between amino acids.
- Exit of tRNA from the ribosome as polypeptide is formed.

Chapter 1.5

1. The process by which a cell develops more specialised functions by expressing the genes characteristic for that type of cell.

2. **a)** meristem cells (1)

   b) stem cells (1)

3. (Red) bone marrow.

4. Provide information on cell processes such as cell growth / differentiation / gene regulation (any 1).

5. Corneal transplants, skin grafts for burns, or treatment for Alzheimer's disease, other (any 2).

6. Embryos must not be allowed to develop beyond 14 days, around the time a blastocyst would be implanted in a uterus during IVF.

Extended response (any 4)

- Use of embryo stem cells destroys embryo, which many people believe is unethical.
- Some believe that the need to relieve suffering overrides all other issues
- Use of induced pluripotent stem cells – not true stem cells, so fewer people have ethical issues.
- Use of nuclear transfer techniques
- – some believe it's unethical to mix human cells and those from another species.
- Others support this as an alternative to embryonic stem cells.
- Some believe this will allow scientists to develop new treatments for diseases.

Chapter 1.6

1. Coding and non-coding.

2. Proteins.

3. tRNA (1), rRNA (1).

4. Exons code for the expression of proteins (1). Introns are removed during splicing and so do not appear in the mature transcript after translation (1).

Extended response (any 5)

- The genome of an organism is its entire hereditary information encoded in DNA.
- DNA sequences that code for protein are defined as genes.
- There are coding and non-coding sequences.
- A genome is made up of genes and other DNA sequences that do not code for proteins.
- Non-coding sequences include those that regulate transcription and those that are transcribed to RNA but are never translated.
- Non-translated forms of RNA include tRNA and rRNA.
- Some non-coding sequences have no known function.
- Most of the eukaryotic genome consists of non-coding sequences.

Chapter 1.7

1. Changes in the genome that can result in no protein or an altered protein being expressed.

2. **a)** A = Deletion, B=Insertion , C= Substitution.

   b) 1 = major, 2 = major, 3 = minor.

3. Mutations create variations in alleles (1), this contributes to the evolution of a species because it creates differences in the survival chances of individuals (1).

4. Insertion or deletion of nucleotides which results in every subsequent codon being different (1); this changes all the amino acids and so synthesises a very different protein (1).

5. The impact will be the change in amino acid sequence, which will in turn affect protein synthesis. This may result in the protein functioning differently or no protein being produced at all.

Extended response

1. - Substitution – a base replaced by another, with no other bases changing. Causes a minor change in protein (1).

   - Insertion – an additional base inserted into the sequence. Causes a major change in protein (1).

   - Deletion – a base deleted from the sequence. Causes a major change in protein (1).

2. - Duplication (1) – produced when extra copies of genes are generated on a chromosome (1).

   - Deletion (1) – this mutation results from the breakage of a chromosome in which the genetic material becomes lost during cell division (1).

- Inversion (1) – a broken chromosome segment is reversed and inserted back into the chromosome (1).

- Translocation (1) – a piece of chromosome detaches from one chromosome and moves to a new position on another chromosome (1).

(Any 6)

3. • Polyploidy occurs when errors during the separation of chromosomes during cell division result in cells with whole genome duplications.

- Polyploids tend to grow to be stronger, larger and more productive.

- Examples include strawberries, bananas and wheat.

- As a result of having additional sets of chromosomes, these organisms have an evolutionary advantage, due to their increased ability to mask harmful recessive alleles.

- In order to inherit a recessive trait, all of the alleles must be recessive.

- Polyploidy organisms require the inheritance of more than two recessive alleles, reducing the odds of showing the harmful characteristic or trait.

(Any 5)

Chapter 1.8

1. The gradual changes to organisms over time due to the effects of natural selection on variation within species.

2. Vertical inheritance – genes passed from parent to offspring (1).

   Horizontal inheritance – genes passed from individual to individual (1).

3. Organisms within a species vary (1).

4. Sexual selection favours characteristics which contribute to breeding success (1).

5. Genetic drift is random changes to the DNA sequence (1).

   The founder effect is the genetic drift seen in small, isolated populations (1).

Extended response

1. • Hybrid zones are regions in which two species interbreed to produce hybrids (1).

   - Hybrid individuals are less fit than their parents (1).

   - Natural selection removes hybrids from the population (1).

   - Zones are maintained for further hybridisation (1).

2. i.
- In allopatric speciation, gene flow between sub-populations of a species is prevented by a physical/geographical barrier.
- Natural selection is different for each sub-population
- Over long periods of time, sub-populations become so different that they can no longer interbreed successfully.

(All 3)

ii.
- In sympatric speciation, gene flow between sub-populations of a species is prevented by an ecological/behavioural barrier
- Barrier arises when individuals acquire mutations which change their ecology or behaviour but don't otherwise affect their survival.
- Further mutations are not shared and natural selection affects the sub-populations differently
- Over long periods of time, sub-populations become so different that they can no longer interbreed successfully.

(Any 3)

Chapter 1.9

1. Sequencing the order of nucleotide bases in an entire genome

2. Bioinformatics

3. Bacteria (1)

Archaea (1)

Eukaryota (1)

4. Prokaryotes, last universal ancestor, photosynthesis, eukaryotes, multicellular organisms

(All = 2, any three in order = 1)

Extended response

1.
- Personalised genomics is having an individual's genome sequenced
- Analysis of an individual genome could lead to personalised medicine.
- Genetic components of disease could be revealed
- The likelihood of success of treatments could be estimated

- There are problems due to the complex multifactorial nature of many diseases
- Neutral mutations that don't cause disease make interpretation of genomes more difficult

(Any 5)

2. • Phylogenetics is the study of evolutionary interrelatedness
   • Sequence data and fossils used
   • Phylogenetic trees can be drawn

(Any 2)

   • A molecular clock diagram compares genetic difference between pairs of related organisms.
   • Relies on rates of mutation being steady.

(Both)

Unit 2

Chapter 2.1

1. Anabolic pathways involve the synthesis of large molecules from smaller/simpler molecules using up energy (1).

   Catabolic pathways involve the breakdown of large molecules into smaller/simpler molecules with the release of energy (1).

2. Avoids excess product(s) by conversion back to original substrate which may be used elsewhere.

3. Allows the control of the rate of an enzyme-catalysed reaction.

4. • Allows isolation of metabolic pathways from areas of the cell where different pathways are taking place.
   • High concentrations of reactants can be maintained.
   • Generates large surface area for reactions to take place.
   • Diffusion of subtrate / product

(Any 1)

5. a) Correctly labelled axis (1); suitable scale (1); correct plot of substance X (1); correct plot of substance Y (1).

   b) When the external concentration is greater than the internal concentration.

   c) Diffusion.

   d) Active transport (1).

      When external concentration is less than internal, the cell uptakes the substance (1).

Extended response (any 6)

- Double layer of phospholipid and protein.
- Phospholipid molecules are in a constant state of motion.
- Protein molecules are randomly scattered within and on the surface of the phospholipid molecules.
- Some proteins are embedded within the double layer of phospholipids.
- Some proteins span both layers.
- Channel proteins have a central pore to allow entry / exit.
- Protein pumps are present in the membrane for active transport.
- Enzymes may be present.

Chapter 2.2

1. Avoids the enzyme pepsin accumulating when it is not needed / makes sure the enzyme pepsin is synthesised only when it is needed.

   Prevents energy wastage generating product which is not required.

   (Any 2)

2. The activation energy would be lowered.

3. A – without inhibitor. As substrate concentration increases, the reaction rate increases and is always greater than B and C (1).

   B – with competitive inhibitor. Rate is lower than A but with increasing substrate concentration, there is an increased chance of the substrate, not the inhibitor, binding to the active site (1).

   C – with non-competitive inhibitor. Rate is lower than A and B and remains low, even if the concentration of substrate increases, because the active sites are blocked (1).

4. a) At 40°C, the total amount of product formed continues to increase as the reaction proceeds for a longer time (1). At 60°C, the total amount of product formed is greater than that at 40°C for the first 6 minutes. It increases as the reaction proceeds for the first 4 minutes (1)

   then quickly levels off after 4 minutes (1).

   No further product is formed after this time (1).

   b) After 5 minutes, the enzyme at 60°C has become denatured and so can no longer operate. No further product is produced (1).

   After 5 minutes, the enzyme at 40°C is continuing to operate and further product is produced (1).

Extended response (any 5)

- Substrate and active site may not bind initially.
- Substrate has affinity for the active site.
- Active site is flexible.
- A change in the active site occurs.
- Induced by the binding of a specific substrate to the active site.
- The enzyme can now cause catalysis.

Chapter 2.3

1. Removes hydrogen from a substrate.
2. FAD and NAD.
3. Adenosine triphosphate/ATP.
4. a) Hydrogen is released much quicker in sample A than sample B.

    b) Use water bath to control temperature; make up the yeast suspension with exactly measured volume of water and mass of yeast; ensure the same kind of containers are used for each sample, such as identical boiling tubes; use an electronic device to measure the colour change rather than using the human eye.

    c) Repeat the experiment many times or run the experiment with multiples of each sample.

    d) Use a third sample with an identically measured volume of boiled/dead yeast.

Extended response (any 6)

- Enzyme complex which catalyses the synthesis of ATP from ADP and P_i.
- Found within the inner membrane of a mitochondrion.
- Also found in membranes of chloroplasts.
- Protein spans the membrane.
- Hydrogen ions flowing through the central channel of the enzyme provide the energy to drive the enzyme.
- Flow of hydrogen ions is down the concentration gradient.
- Acts as a rotary engine which, as it rotates, alters its active site.

Chapter 2.4

1. Within the matrix/within the inner membrane of the mitochondria.
2. Picks up hydrogen produced in cell respiration (1);

   passes this to the electron transport chain (1).

3. Lactic acid (1);

   ATP (1).

4. Oxygen.

5. a) To maintain the temperature at the optimum level for each tube.

   b) Mass of beans in each tube; type of beans in each tube; temperature throughout the experiment; diameter of the boiling tubes; mass of soda lime in each tube (Any 3).

   c) Same mass of same beans but previously killed by boiling, for example.

   d) Note the level of the coloured liquid in the capillary tube at the start of the experiment; record the change in the height of the column of coloured liquid at regular intervals of time.

Extended response (any 5)

- Energy-investment phase results from initial reactions.
- 2 ATP are used up.
- Provides energy to drive subsequent reactions.
- Energy pay-off phase results from later reactions.
- Phosphorylated intermediate compound is converted to 2 pyruvate molecules.
- 4 ATP are produced.
- Overall gain is therefore 2 ATP.

Chapter 2.5

1. Measuring: oxygen consumption, carbon dioxide production and energy production per unit time (All 3 = 1 each).

2. a) **False:** A fish has a heart with **one ventricle** with the blood flowing from the **heart** to the **gills** (1).

   b) **False:** Amphibia and **reptiles** share a similar heart structure with blood flowing through the heart **twice** for each circuit of the body (1).

   c) **False:** Birds have a heart which is divided into **two atria** and **two** ventricles with blood under **high** pressure (1).

   d) **False:** The skin of a frog must be **wet/moist** to allow gas exchange to occur (1).

e) **False:** The lungs of a mammal have **many** air sacs which generate a **large** surface area for gas exchange (1).

3. **a)** Correct labels, units and scale for each axis (1); correct plotting (1).

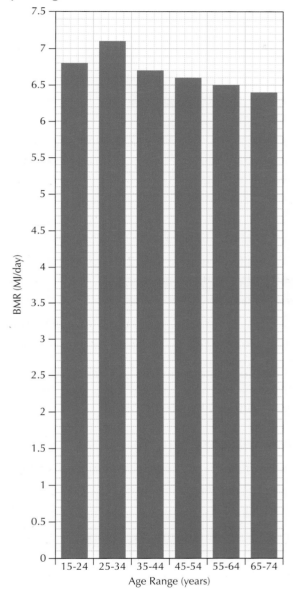

b) In general, BMR decreases with age; the highest BMRs are found in 25–34 year olds.

c) Change is $7 \cdot 1 - 6 \cdot 8 = 0 \cdot 3$; % change is $(0 \cdot 3 \div 6 \cdot 8) \times 100 = 4 \cdot 4\%$.

Extended response (any 6)

- Can do without normal breathing for over an hour.
- High concentrations of oxygen-carrying pigments in muscles in addition to haemoglobin.
- Large volume of blood.
- Many red blood cells.
- Can reduce rate of heartbeat considerably during diving.
- Lungs collapse on diving to avoid 'bends'.
- Relies on stored oxygen for aerobic respiration during diving.
- Precise regulation of bloodflow to where it is needed.
- Conserve energy by using gliding motions instead of active swimming.

Chapter 2.6

1. Salinity, pH and temperature (3 = 2 / 2 or 1 = 1).
2. All correct for 2 marks; 3 or 2 correct for 1 mark.

Conformers	Regulators
A, B	C, D

3. a) Conformer (1).

   Unable to maintain salt concentration of body fluid/ mirrors changes in the salt concentration of the environment (1).

Extended response (any 6)

- A mechanism for achieving homeostasis.
- Keeps internal environment steady.
- When factor changes this is detected by a receptor.
- Stimulates mechanisms to reverse the change using effectors.
- Via effector(s).
- Restoring set point value.
- Relies on a combination of nervous and hormonal signals.
- Once corrective mechanism has restored set point value, the strength of the response decreases.

Chapter 2.7

1. **a)** Predictive (1). Dormancy occurs before the adverse conditions develop (1).

    b) Shortening daylength/photoperiod; decreasing temperature.

2. a) and c)

3. **a)** Suitable scales which use more than 50% of the grid (1); labelling each axis and including units (1); plotting data (1 for each set).

    b) State 1 is arousal. State 2 is torpor.

    c) At any given temperature, the oxygen uptake of the hummingbird in a state of torpor is always less than the hummingbird in a state of arousal (1). For both states, the oxygen uptake decreases as the air temperature increases (1).

Extended response (any 4)

- An organism which thrives in an environment whose conditions are hostile to most forms of life on the planet.
- Such conditions include extremes of temperature, pH, salt concentrations, atmospheric air pressure and radiation levels.
- Most extremophiles are found in the domain archaea.
- Some bacteria can thrive in deep-sea vents where the water temperature can exceed 80°C.
- Not only are such environments exceptionally hot, they are also very rich in gases which may be toxic to most forms of life.
- These gases include methane and sulfur compounds which are actually used by the bacteria as energy substrates.

Chapter 2.8

1. Culturing

2. **a)** 3,500,000

    b) 24 minutes

    c) 5

3. **a)** C

    b) D

    c) B

4. 3

Extended response (any 6)

- Secondary metabolism in microorganisms produces metabolites of potential use to humans.
- Pathways associated with secondary metabolism can be manipulated in such a way as to make the microorganism produce more of a desired metabolite than it would do naturally.
- Secondary metabolites include antibiotics / chemicals for treatment of tumours or lowering cholesterol levels / pigments.
- Precursors form more complex compounds in the next stage of a metabolic pathway.
- Addition of precursors will ensure the formation of the next compound in line is maintained and/or enhanced.
- Inhibitors and inducers target enzyme function by stopping or speeding up the production of that enzyme.
- This in turn will prevent the next metabolite being formed or ensure the next metabolite continues to be made respectively.

Chapter 2.9

1. Chemically: asbestos/mustard gas (1). Physically: radiation (1).
2. Microorganism does not express some unwanted trait or metabolite/produces an increased yield of a wanted metabolite.
3. D.
4. Introducing genes which stop the microorganism being able to reproduce in the external environment/aseptic techniques.
5. Gene for antibiotic resistance/gene causing fluorescence (glowing) under special light source.

Extended response (any 6)

- Isolate human gene for insulin production from donor chromosome.
- Using PCR, make multiple copies of this gene.
- Insert genes into bacterial plasmids.
- Use endonuclease to remove human gene and same endonuclease to open up plasmid.
- Use ligase to seal gene into plasmid.
- Add marker to identify which bacteria have picked up the modified plasmids.

- Culture large quantities of genetically engineered bacteria which have insulin producing gene.
- Harvest insulin.

Chapter 2.10

1. Biotechnology allows scientists to manipulate the genetic material found in microorganisms. However, such procedures can raise ethical issues associated with potential hazards. Since microorganisms can multiply easily and rapidly and are highly adaptable, there is the danger of them escaping into a new environment and colonising it. Because microorganisms are haploid, any mutations will be expressed and if these confer, for example, increased virulence or resistance to antibiotics, there is a danger to other species, including humans. (All 11 correct for 5 marks/9 or 10 correct for 4 marks/7 or 8 correct for 3 marks/5 or 6 correct for 2 marks.)

2. Designing safe buildings, customised for this type of work (e.g. air-tight/negative pressure/filtration); ensuring proper aseptic protocols are rigidly adhered to; using effective chemical sterilising agents; using physical agents such as extreme heat/radiation to sterilise equipment before and after use; disposing of all cultures safely; genetically engineer microorganisms so they are unable to survive outside laboratory conditions of growth (Any 3).

Unit 3

Chapter 3.1

1. a) People have food security if they are able to access enough high quality food at all times, can afford to buy it and have the knowledge to use it properly.

    (All = 2; any 2/3 = 1).

    b) Increase in the size of the human population (1); increase in money available to spend on food (1).

    c) Increasing crop yields while at the same time, conserving natural resources.

    d) i. Less competition for water, space and nutrients.

    ii. Application of nitrogen fertiliser increases the growth rate of plants, resulting in an increase in dry mass/productivity of crop.

2. a) i. 2011 barley yield = 1·4 million tonnes

    2012 barley yield = 1·7 million tonnes

    Increase in yield = 0·3 million tonnes

Percentage increase = 0·3 ÷ 1·4 (initial yield) × 100 = **21.4%**

ii. 2011 wheat yield = 2·1 million tonnes

2012 wheat yield = 1·3 million tonnes

Decrease in yield = 0·8

Percentage decrease = 0·8 ÷ 2·1 (initial yield) × 100 = **38%**

iii. Average total yield of both cereals over the two year period.

1·4 + 1·7 + 2·1 + 1·3 = 6·5 million tonnes

6·5 ÷ 4 (total number of numerical values added together) = **1.625 million tonnes**

b) Axes scaled and labelled = 1. Plotting and line connection = 1. Points plotted accurately using appropriate number scales = 1 mark. Both axes labelled clearly including units of measurement = 2 marks.

Extended response (any 6)

- Efficient irrigation methods.
- Improved soil management practices
- Accurate stocking density per hectare of land
- Optimum levels of fertiliser application to maximise both plant growth and productivity.
- Use of organic, low energy fertilisers as opposed to chemicals.
- Use of crop rotation systems rather than monoculture.
- Minimal use of fossil fuels to power machinery.
- Use of high yielding cultivars.

Chapter 3.2

1. a) Absorbed, transmitted, reflected (3 = 2; 2 or 1 = 1)

b) Absorption spectrum: wavelengths/colours of light absorbed by each pigment. Action spectrum: effect of each wavelength of light on photosynthesis in terms of product.

c) NADPH and ATP needed to convert 3-phosphoglycerate to G3P.

d) Extended response (any 6)

- Photosynthetic pigments are located in the chloroplasts of plants

- Main pigments are chlorophyll a and chlorophyll b
- Pigments absorb light energy
- Light energy absorbed by chlorophyll a excites electrons into a high energy state
- High energy electrons pass through the electron transport chain producing hydrogen ions
- Hydrogen ions rotate ATP synthase phosphorylating ADP + Pi to ATP
- Some light energy absorbed by pigments is used to split a molecule of water into hydrogen and oxygen
- This process is called photolysis
- ATP and NADPH are produced for the Calvin cycle

2. a) **i.** Plant Y – xanthophyll – 0·4 µg/cm^3, carotene – 0·67µg /cm^3

Average mass of carotenoid pigments = 0·49 + 0·67 = 1·16

1·16 ÷ 2 = **0·58µg/cm^3**

ii. Carotene (plant X)   xanthophyll (plant Y)

                    0·52   0·26

Ratio   **2**   :   **1** (0·52 is twice 0·26)

iii. Plant Y – mass of chlorophyll a = 0·94µg/cm^3

Plant X – mass of chlorophyll a = 0·90µg/cm^3

Increase in mass of chlorophyll a = 0·04µg/cm^3

% increase in mass = 0·04 ÷ 0·90 (initial value) x 100 = 4.444%

iv. Plant X – xanthophyll=0·26µg/cm^3, carotene = 0·52µg/cm^3

Plant Y – xanthophyll = 0·49µg/cm^3, carotene = 0·67 µg/cm^3

Average mass of accessory pigments –

Plant X = 0·26 + 0·52 = 0·78µg/cm ÷ 2 = 0·39µg/cm^3

Plant Y = 0·49 + 0·67 = 1·16 ÷ 2 = 0·58µg/cm^3

Plant Y contains a higher average mass of carotenoid pigments = 0·58µg/cm^3 compared to plant X = 0·39µg/cm^3, and so absorbs more light energy.

    b) **i.** Scales and labels on axes (1); plotting of graph (1).

        ii. 25°C maximum volume of oxygen = 3·5cm^3

        35°C maximum volume of oxygen = 7·4 cm^3

        Volume of oxygen produced indicates less activity of photosynthetic enzymes at 25°C, and increased activity at 30°C as this temperature is nearer to the optimum temperature for enzyme activity.

      iii. Photosynthetic enzymes had become denatured.

      iv. 6 minutes.

      v. 35°C – initial volume of oxygen = 4·1 cm^3, final volume after 20 minutes = 7·4 cm^3

      Increase = 3·3 cm^3

      % increase = 3·3 ÷ 4·1 × 100 = **80·47%**.

Chapter 3.3

1. a) True.

   b) False – increase in carbon dioxide leads to an increase in biomass due to an increase in photosynthesis.

   c) False – net assimilation + increase in biomass – decrease in biomass due to cell respiration.

   d) True.

   e) False – increased distance between young plants would result in increased areas of bare soils and reduced leaf area index.

   f) True.

Extended response (any 6)

- Increase in carbon dioxide concentration results in an increase in the rate of photosynthesis unless other factors are limiting.
- Biomass would increase.
- Increase also in net assimilation.
- More carbohydrate would be converted into cellular structures.
- Crop productivity would increase.
- There would be an increase in the greenhouse effect.
- This would cause an increase in global temperature.
- Available soil water would decrease.

2. a)

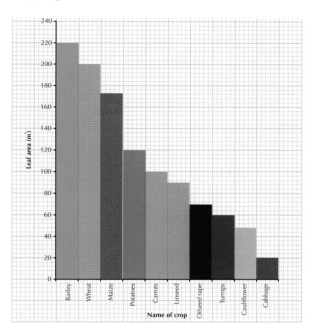

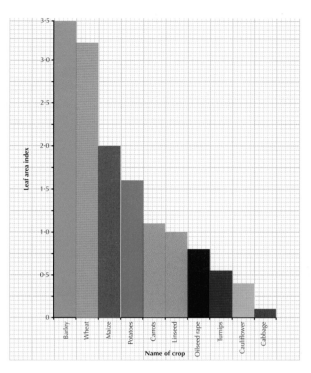

For each graph: Accurate plot = (1); accurate naming of axes (1); sensible scale (1)

b) Potatoes – leaf area = 120m², Cauliflower – leaf area = 48m²

Decrease in leaf area = 72m²

% decrease = 72 ÷ 120 × 100 = **60%**

c) Leaf area is proportional to leaf area index. As the leaf area increases, the leaf area index increases.

Chapter 3.4

1. Animals – increased milk yield, increased body mass, increased egg size, disease resistance. Plants – increased yield/grain size, increased resistance to disease and drought. (Any 4 = 1 each)

2. Inbreeding

3. a) F_1 offspring all Bb (1), all have black coat (1).

b) F1 offspring crossed together Bb × Bb

Gametes   B   b

           B   BB   Bb

           b   Bb   bb

Phenotypic ratio = 3 black coat calves to 1 white. (1)

Extended response (any 6)

- A field trial is carried out.
- A number of plots are sown with seeds of the new barley cultivar.
- Plots are all the same area.
- Barley is grown in randomised plots across the whole experimental field area.
- All plots are treated the same to ensure a fair comparison.
- Crop is allowed to grow to maturity.
- Crop from each plot is harvested.
- Data is recorded such as dry mass, leaf area and economic yield.
- Controlled by same trial with the old cultivar

Chapter 3.5

1. Annual weeds – sexual reproduction, many seeds produced, short lifecycle, rapid growth. Perennial weeds – asexual reproduction, no seeds produced, long lifecycle, slow growth, storage organs only. (Any 3 = 3; 2 or 1 = 1)

2. Physical removal of weeds by hand, ploughing, cultivating between rows of plants.

3. Contact herbicide – non-selective, kills all plants on contact of chemical with leaves. Systemic herbicide – non-selective, absorbed into xylem and phloem vessels, kills all parts of plant.

4. Insects, nematodes and molluscs (1 each).

5. a) i. Pig meat imports – October 2013 = 79,514, October 2014 = 76,461

   Decrease = 3,053 tonnes

   % decrease = 3053 ÷ 79514 (initial value) × 100 = 3·8%

ii.

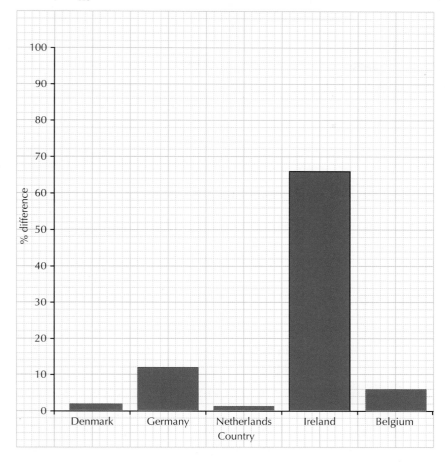

Graph = 3

All 5 = 3; 4 or 3 = 2; 2 or 1 = 1

b)

Year 2	
1 BARLEY	2 POTATOES
3 TURNIPS	4 GRASS

Year 3	
1 TURNIPS	2 BARLEY
2 GRASS	4 POTATOES

Year 4	
1 GRASS	2 TURNIPS
3 POTATOES	4 BARLEY

Extended response (any 6)

Advantages:

- Negative impact on the environment is reduced.
- A natural predator is used to reduce the size of the pest population.
- Fewer chemical pesticides are needed
- Example – greenfly in glass houses can be controlled through the introduction of ladybirds – a natural predator of greenfly.

Disadvantages

- Pest population size is reduced, but not eliminated completely.
- Introduced predator may in turn become a pest.
- Balance of food chains and food webs can become upset within local ecosystem.

Chapter 3.6

1. **a)** Two different species which have co-evolved and live together in a close relationship.

    b) Parasitic relationship: one species benefits from available resources while the other is harmed.

    Mutualistic relationship: both species benefit from available resources.

2. • Increased chance of finding larger prey (1).

   • Reduced competition between members of the group (1).

   • Each member of the group receives a share of the kill, each individual gains more energy (1).

3. • By giving priority to family members (looking after relatives), in order to protect shared genes (1).

   • Individuals may sacrifice their own lives in order to preserve the lives of other family members, preserving shared genes for future generations (1).

4. **a)** **i.** 523×0.32 (32%) = 167 geese killed

   ii. • 75% of males were found on the perimeter of the flock compared to 25% females.

   • Less time is spent scanning horizon for predators – males 7·8% of time spent scanning horizon compared to 10·4% in female geese.

   iii.

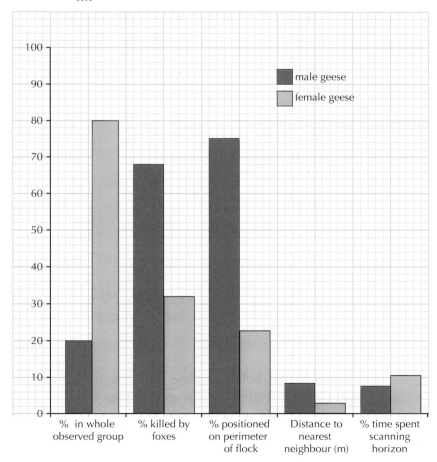

Extended response 1 (any 6)

- Only the queen bee lays eggs.
- Other female bees are called workers and are sterile.
- Males fertilise the eggs of the queen bee.
- Drone bees care for young larvae, provide food for young larvae and the queen bee.
- Drones keep breeding cells (nursery area) clean.
- Drones defend the hive.

Extended response 2 (any 10)

- High degree of parental care.
- Young learn social skills through play (seen in young chimpanzees).
- Social hierarchy structure exists within a group of primates.
- Each individual has a defined position within the group, with the most dominant individual at the top of the hierarchy.
- Hierarchical structure reduces conflict and conserves energy.
- Threat displays reinforce individual positions within the hierarchy.
- Examples – walking upright, baring teeth, dragging branches.
- Submissive individuals display appeasement behaviours.
- Examples – crouching, grooming.
- Alliances form between individuals.
- Alliances help reinforce individual position within a social group.
- Opportunity for individuals to progress upwards in hierarchy through formation of alliances.
- Social structure influenced by external factors – ecological niche, resource distribution, taxonomic group.
- Ecological niche is the role of an individual within the environment.
- Primates which belong to the same taxonomic group share similar social structures and ecological niches.

Chapter 3.7

1. **a)** Catastrophic environmental change caused by, e.g., a meteor collision with Earth, resulting in a significant decrease in biodiversity in a short period.

   b) Human activity such as hunting and reducing their habitats by agricultural practices and clearing land for roads and cities. Industrial pollution of land, air and sea disrupting the food chain as well as rendering habitats uninhabitable. They are slow breeders, are seen as competition for resources, are seen as predators, need lots of habitat. (Any 3 = 1)

   c) The extinction rate is the total number of species which have become extinct within a specific geographical area over a specific period of time (1). Calculating an estimated value for extinction rate involves observing and recording numbers of animals and plants within a specific area (1).

   d) Species diversity (1); genetic diversity (1); ecosystem diversity (1).

2. **a)**   **i.** 20 years

       ii. Overfishing – removing young cod before they are mature enough to breed. Fishing of associated species such as haddock, whiting, plaice.

       iii. 30,000–40,000 tonnes per week. Another 30,000 tonnes is needed to stabilise the cod population.

       iv. Population size reduced, fish stocks will collapse due to large numbers of fish being removed from population before they reach maturity and are able to reproduce.

   b)   **i.** Terrestrial ecosystem.

       ii. 2008.

       iii. 1960: number of species = 4; 1980: number of species = 6. Increase = 2.

       Percentage increase = $2 \div 4 \times 100 =$ **50%**.

       iv. Invasive species able to travel more easily over land. Wider range of habitats for invasive species to colonise.

Extended response (any 8)

- Human activities such as felling trees to make roads, ploughing land for crops, building houses, etc., can break habitats into fragments or 'islands'.
- Some populations become divided and isolated as a result.
- Species diversity and gene pool are reduced due to smaller number of individuals in sub-population.
- Large fragments become broken up into smaller fragments through further human activities such as new roads.
- Fragment edges suffer erosion due to, e.g., increased exposure to wind and rain, decreasing size of habitat area.
- Habitat corridors can be natural or man-made.
- They connect habitat fragments together.
- Individuals can now move between isolated populations.
- Biodiversity restored as individuals can now outbreed.
- Populations stabilised.
- Example – tunnels built under motorways allow hedgehogs, badgers and small mammals to move safely between isolated habitat islands, avoiding being killed by cars, on either side of the road.